AF683967

Clinical Chaos

Marie-Louise von Franz, Honorary Patron

Studies in Jungian Psychology by Jungian Analysts

Daryl Sharp, General Editor

CLINICAL CHAOS

The Strange Attractors of Childhood Trauma

John R. Van Eenwyk

INNER CITY BOOKS

For Juliet

Library and Archives Canada Cataloguing in Publication

Van Eenwyk, John R. (John Richter), 1946-
Clinical chaos : the strange attractors of childhood trauma / John R. Van Eenwyk.

(Studies in Jungian psychology by Jungian analysts; 135)
Includes bibliographical references and index.
ISBN 978-1-894574-37-2

1. Psychic trauma. 2. Psychic trauma—Case studies. 3. Jungian psychology. 4. Chaotic behavior in systems—Psychological aspects. I. Title. II. Series: Studies in Jungian psychology by Jungian analysts ; 135

BF175.5.P75V35 2013 150.19'54 C2012-904952-2

Copyright © 2013 by John R Van Eenwyk.
All rights reserved.

INNER CITY BOOKS
Box 1271, Station Q, Toronto, ON M4T 2P4, Canada.
Telephone (416) 927-0355 / Fax (416) 924-1814
Toll-free (Canada and U.S.): Tel. 1-888-927-0355 / Fax 1-888-924-1814
Web site: www.innercitybooks.net
E-mail: booksales@innercitybooks.net

Honorary Patron: Marie-Louise von Franz.
Publisher and General Editor: Daryl Sharp.
Associate Editor: Frith Luton.
Senior Editor: Victoria B. Cowan.
Office Manager: Scott Milligen.
Technical Support: David Sharp (www.sharpconnections.com)

INNER CITY BOOKS was founded in 1980 to promote the understanding and practical application of the work of C. G. Jung.

Cover Image: Fractal image of galaxy M101 (see p. 52).

CONTENTS

PREFACE & ACKNOWLEDGMENTS

In 1997 I published *Archetypes & Strange Attractors: The Chaotic World of Symbols,* a theoretical study of the relationship between Analytical Psychology and chaos theory in physics.[1] Surprisingly well received, it continues to find an audience today. As a result, I have been consistently asked how it applies to clinical challenges. What lessons can we take into the consulting room? That is the heart of this sequel.

For answers, I had to rely on my patients. Without them, the present volume would be little more than speculation. Thus, I owe everything in this book to the people who have found their way to my door over the last forty years.

They have taught me a great deal. They have been wonderful colleagues and fellow researchers. It has been an honor to work with them.

I also owe a great deal to my mentors. There are the academics who were wonderfully patient as I struggled to understand what they were teaching: Warren Ramshaw and Tony Aveni at Colgate University, Harvey Guthrie and Joseph Fletcher at The Episcopal Theological School, Norman Watt and Arthur Dyck at Harvard University, Mircea Eliade and Bernie McGinn at the University of Chicago, Ian Baker and Marie-Louise von Franz at the C. G. Jung-Institut Zurich. There were also great spiritual leaders who helped me to engage the ultimate mystery: Richard Cohoon, Don Stivers, and George Barrett. I am grateful to Sydney Harris for permission to use his cartoon on page 57.

I want also to thank Daryl Sharp for publishing my work and that of many other Jungian analysts. His small publishing house does a great service to all of us. I thank him and Frith Luton for their editorial work.

Finally, the reader. What would be the use of any of this without you? Thank you for being part of the research. Please take up where this book leaves off. And, please, do let me know how it goes, c/o Inner City Books, or email: jrv@u.w.edu

---

[1] While it would be helpful for readers to be familiar with the concepts presented in that volume, the understanding of *Clinical Chaos* does not depend on it.

INTRODUCTION

The word *chaos* denotes a phenomenon immediately recognizable but notoriously difficult to define. It has much in common with *in*-words like ineffable, inscrutable, incomprehensible, indescribable, indeterminate, and inconceivable that rarely get at what it is; only at what is not. So in the first volume of this series, *Archetypes & Strange Attractors: The Chaotic World of Symbols* (hereafter referred to as *Archetypes).* I described the place that chaos holds in Jung's metapsychology.

The emerging field of *chaos theory* provided a basis for that comparison. Through mathematical equations and computer-generated graphs, chaos theory has done much to unlock the characteristics of complex dynamics. That made it possible in *Archetypes* to use chaos theory to illuminate Jung's structure and dynamics of the psyche and Jung's theories to clarify complex dynamics.

As my interest in both Jung and chaos theory originated in the consulting room, eventually I had to relate it to clinical experience. That took a while. Finding the right examples was more difficult than I had imagined. As with all case material, I had to be especially careful to reveal as little as possible about each patient. That, too, took time.

My use of the word *patient* may cut across the grain. I realize it has some negative connotations, not the least of which is the pigeonholing of perfectly functional people into categories labeled *impaired*, *deficient*, or worse, *disordered*. But I believe that the solution lies less in expunging the term than in using it to describe all of us.

Moreover, sorting through the chaos of the psyche requires a lot of patience. I have been fortunate in my career to be able to work with those who are wonderfully patient with my lack of understanding, knowledge, and empathy. With a great deal of forbearance we have explored together the inscrutable recesses of the mind.

I don't like the word *client*. While I make my living doing analysis, patients are not purchasing something from me. They are simply making it possible for me to keep house and home together while we go on the journey together. Without that support, I could not accompany them.

Nor do I like the expression *consumer of mental health services*. That's the kind of apotropaic euphemism that encourages diarrhea of the brain. Even the word *analysand* seems unduly ponderous. So, for better or for worse, I use the word *patient*.

The Clinical Situation

Clinicians routinely sense that something is going on that resists definition. We're trained to believe that certain symptoms indicate certain etiologies. But what if there is no evidence of cause? How are we to make sense of things if we have nothing to go on?

Furthermore, what constitutes healing when the disease is unknown?

We won't be far off the mark if we assume that traumatic experiences permeate many psychological difficulties. Firstly, everyone is traumatized, some more so than others. But we all react to trauma differently. Some endure the most horrific of traumas with little scarring. Others are devastated by what to some may seem merely incidental. Whatever the reaction, trauma is a part of life.

Secondly, the symptoms of trauma have been pretty well discovered and described. We're not as much in the dark as we are with anxiety or depression. Jung's belief that symptoms are attempts at cures also helps us to know how to get started.

Finally, the primary effect of trauma is disconnection, from self and from other. Thus, the analytic relationship is healing in itself, assuming that the analyst is sufficiently aware of her or his own psychological idiosyncracies to avoid burdening the patient with them. Thus, the treatment for trauma is consistent with the analytic process.

This, then, is a book about trauma. We'll concern ourselves less with the single-event kind, than with the chronic trauma that can permeate early life experience. Such patients who enter analysis later in life usually do not remember the traumatic environment in which they were raised. Rarely do they think of themselves as traumatized. Rather, they seek analysis because life just isn't going as it should. They are unhappy, lonely, isolated, unfulfilled, frustrated, angry, depressed, confused. What they all have in common is that they don't know why.

They hope the analyst does. And in fairness to them, that is what we are paid to do.

But the answers lie with our patients. Only they hold the information we need to determine what happened, how their psychic apparatus adapted to it, and what we can do about it now. Consequently, analysis is a mutual endeavor. Analyst and patient work together to discern, understand, and transform the sequelae of early childhood trauma.

Abusive and neglectful environments rarely change as children grow older. What does change, however, is how children cope. Until their cognitive capacities are developed sufficiently to sort everything out and adopt coping strategies, the unconscious does it for them. It all happens automatically. Later on, however, cognitive capacities construct an overlay onto the unconscious coping strategies. That leads to a split in the psyche between conscious and unconscious coping strategies.

Because unconscious coping strategies develop early in life, they involve the deeper levels of the brain, primarily the amigdala, where the lateral nucleus encodes the unconscious's adaptations in memory. Later in life, conscious strategies, which are a function more of the prefrontal cortex, find themselves in conflict with the unconscious strategies, to the discomfort of the individual. Cognition and emotion must somehow be reconciled.

The Strategy

That is the focus of this book. How can we "read" the unconscious to discern the presence and dynamics of its coping strategies? Once these are discerned, how can we transform them? We cannot do it solely through consciousness, for while rationality is an excellent way to sort things out—through description, definition, categorization, and logic—the unconscious simply transcends those capacities. For example, whenever we define or describe, we leave out more than we include.

Consequently, the goal of analysis is to integrate conscious and unconscious.[2] As we shall see, that involves bringing unconscious into conscious and conscious into unconscious. The former usually feels easier than the latter. But the latter is equally, if not more, important to

[2] Excuse my syntactic imprecision with regard to the use of the words *conscious* and *unconscious*. Sometimes I use them as adjectives, sometimes as nouns (as I do here).

healing.

The question that this book addresses is this: How can we heal the trauma that complicated childhood when the adult has no conscious memory of the trauma? Part One provides a theoretical grounding for Part Two, which concerns itself with treatment. Both describe *morphogenesis*, a term that describes how the psyche forms new patterns out of old. Throughout, we'll be referring to chaotic dynamics and how to work with them in psychotherapy.

The Challenge

Paradox hovers over the clinic. For example, explaining clinical examples of chaos theory from the perspective of Analytical Psychology requires both a reductive/simplifying approach and a constructive/complicating one. That is, to define terms and to describe dynamics we'll have to reduce very complex processes to their simplest forms—their component parts, as it were.

Nevertheless, to do justice to the complexity of the dynamics, we must also elaborate them through rhetorical devices that can be rather abstruse, such as metaphors, allusions and analogies. That such poetics can sometimes seem a bit fuzzy should not be viewed as problematic, for imprecision is often more faithful to complex phenomena than is the precisely specifiable.

Consider the word *psychology*, whose literal meaning is "the study of the psyche." Remarkably, after several thousand years of philosophy and a couple of hundred years of psychology, the word *psyche* remains essentially undefined—if not indefinable. Consequently, to study the psyche or to engage in psychotherapy requires that we be capable of operating without defining exactly what we are studying or engaging.

Not everyone is comfortable with that. Some try to manufacture legitimacy through phrases like *behavioral science*. Others allege that psychotherapy can be *evidence-based* by applying (presumed) scientific analyses to theories and techniques in hopes of isolating those that can confidently be expected to produce the intended results. By applying the scientific method to psychotherapy, it is thought, we can discover what works and what doesn't.

It's a worthy effort. But it has its shadow side. Behind those efforts lie two hugely powerful factors. The first is validity. Just as medicine labored hard and long to extricate itself from superstition and witchcraft, now does psychology seek to do the same. Are we nothing more than charlatans who proclaim our validity through anecdotal evidence that relies for its veracity on our word alone? Or is there objective verification for our theories and practices?

The second factor may be even more powerful than the first. Who pays for our labors? If it's our patients, then they can decide for themselves if our efforts are worth the price. But when third-party payers enter the picture, the margins narrow. Public and private insurers would like to know if they are purchasing anything real. Governments are accountable to taxpayers. Insurance companies are accountable to shareholders—or to the lifestyles of their CEOs.

Those two factors lead psychologists to feel compelled to justify their existence. No longer are credentials and experience sufficient to ensure competence. Now, the demand is for proof. What works and what doesn't? Who decides, practitioner or payer? And on what basis?

Thus is born the quest for certainty, which in today's market is defined as treatments that cure psychological disorders. In other words, which treatments cure which diagnoses. From there, it is but a short step to the Holy Grail of all medicine, namely, that each diagnosis has a treatment that has been reliably shown to cure the disorder. And so we come to psychology today, where depression is cured by medication in the same way that appendicitis is cured by surgery.

Yet, psychotherapy is a bit more complex than medicine. One appendix is pretty much like every other. But depressions? Anxieties? What if someone is completely disheartened by what feels like a meaningless existence? What then? Will it ever be possible to come up with psychotherapeutic techniques that can apply to everyone?

Carl Jung didn't think so. Fifty years ago he wrote that "the more a theory lays claim to universal validity, the less capable it is of doing justice to the individual facts."[3]

---

[3] "The Undiscovered Self," *Civilization in Transition,* CW 10, par. 493. (CW refers throughout to *The Collected Works of C. G. Jung)*

Consequently, he advised, "Generally speaking, the less the psychotherapist knows in advance, the better the chances for the treatment. Nothing is more deleterious than a routine understanding of everything."[4]

Either he was way off, a product of another, less rigorous, less scientific age, or terms like "behavioral science" and "evidence-based psychotherapy" are meaningless—or worse. Perhaps both are correct. Maybe it is possible to study the psyche scientifically. Maybe psychotherapy can be improved by seeking evidence of its results. Or maybe whatever is discovered thereby will be nullified by the next patient who enters the consulting room.

The latter is familiar to most psychologists. Bachelor's degree programs introduce us to a wide range of fields, from anthropology to zoology. For our Master's degrees, we select one of those fields and learn how to use its tools. For our Doctoral degrees we learn how to make the tools—the theories that inform their design; the ways in which experience contributes to their refinement; and the controversies, conflicts and resolutions that accompany their development. Finally, to prove we have learned our lessons well, we are charged with designing a new tool or making a successful improvement on one that already exists.

At the completion of my Doctoral degree and Analytic training, I understood that every patient requires the design and implementation of new and unique tools. For more than forty years, no diagnosis has ever appeared in my office more than once.[5] Subjective individuality trumps objective collectivity every time. Truly, the one thing we all have in common is that each of us is unique.

With all that in mind, I do believe there are shared human experiences to which we react idiosyncratically. A deeper and more complete understanding of such common experiences may allow us to tailor our therapeutic tools to the unique dilemma of each individual. Nevertheless, improvements in psychotherapy will always be asymptotic: We may get

---

[4] "Medicine and Psychotherapy," *The Practice of Psychotherapy,* CW 16, par. 197.

[5] By "diagnosis" I mean a description of the state-of-mind (or state-of-psyche) of the individual. I am not referring to the DSM *(Diagnostic and Statistical Manual of Psychiatric Disorders),* which limits the six billion plus population of the world to a handful of diagnoses.

closer to the ideal, but we'll never actually get there.

So, let's give it a shot. Let's take a hard look at the chaotic dynamics of psychotherapy. By identifying and analyzing dysfunctional elements in our patients' relationships with both themselves and others, we may be able to trace them back to functional responses to disordered situations. We may then be able to understand how those dysfunctional elements derive from responses that made it possible for patients to survive those situations in the first place. In true paradoxical fashion, a brilliant solution to an insolvable problem may now itself have become the problem.

Where Jung and Chaos Theory Meet

That will land us squarely in the realm of chaos theory. Complexity and paradox form the heart of what we shall attempt to understand. As I outlined in *Archetypes*, complex dynamics, explored and described in chaos theory, are the core of Jung's approach to the psyche (however one wishes to define that term).

Thus, this work takes up where *Archetypes* left off. Having explained how chaos theory and Jung's metapsychology reflect and comment upon one another, I then provide clinical examples of how those insights might be applied in the consulting room. Part One is theoretical, to provide a foundation for the methods outlined in Part Two and to encourage readers to form their own conclusions.

Chapter One views the relationship between psychology and physiology through the lens of morphogenesis. Chapter Two outlines how archetypal dynamics react to toxic childhood environments. Chapter Three reviews chaos theory as it applies to clinical work. Chapter Four looks at the paradox of defenses. Chapter Five brings our theoretical elaborations to a close by locating them in trauma theory.

Part Two consists of case studies that illustrate how to apply chaos theory to healing trauma. Chapter Six suggests a four-part scaffold by which to identify the paradoxes of defenses. Chapter Seven illustrates how to reconnect ourselves with those aspects of our psyche that we left behind in childhood. Chapter Eight looks at ways the unconscious provides clues about how to cope with autoimmune disorders and the trauma they beget. Then, Chapter Nine explains how active imagination can become part of our everyday lives, while Chapter Ten reviews and points the way forward.

Caveat

The introduction to *Archetypes* described some of the material it presnted as "chewy," because psychology is never simple. Analysis is not rocket science; its orders of magnitude more difficult. We are all extremely complex. To boil everything down to a few easily grasped essentials is to demean the magnificence of the human spirit, for we are so much more than language can describe.

So, I urge you to stick with it. I'll try to be as clear as possible. If after reading this book and pondering its ideas you find yourself wondering how you could possibly be so clueless, consider the fault mine. You aren't clueless. I am simply inadequate to the task. Together, however, we just might come up with something that can be of use.

That's the best we can do.

PART ONE
Psychology and Theory

Never do human beings speculate more, or have more opinions, than about things which they do not understand.

—C. G. Jung.

1
Psychoid Chaos
The Event Horizon of the Mind

What are the strange attractors of childhood trauma? For that matter, what are strange attractors? In *Archetypes,* the first volume of this series, we saw how a particular form of physical dynamics parallels Jung's accounts of psychological dynamics. Deterministic chaos, which develops when oscillating systems produce cascades, preserve and periodically reproduce the original patterns that got them started.

That is pretty close to Jung's descriptions of the activity of archetypes in psychological development. As the psychological component of instincts, archetypes are the mechanism by which instincts recognize that which helps and hinders growth. Throughout life, archetypes guide development. As they accomplish the task, certain patterns in life are preserved and periodically re-emerge.[6]

These patterns influence our perceptions and behavior. Nevertheless, they are rarely compatible with conscious rationality. That they attract psychological dynamics to certain patterns, yet exceed the powers of conscious rationality to understand them, likens them to the strange attractors of deterministic chaos, for archetypes arrange psychological dynamics into configurations that endure throughout the subsequent chaos of life.

We are most concerned with those configurations that become encoded in the brain before consciousness is formed sufficiently to develop them itself. That these are a function of the unconscious makes them extremely difficult for consciousness to figure them out. Thus, they are *strange* in the same sense as *strange attractors* in deterministic chaos.

Archetypal configurations that form before consciousness is up and running rarely reveal the original experiences that gave rise to them. Thus, it is not unusual for patients to appear in the consulting room presenting with symptoms of Post-Traumatic Stress Disorder (PTSD),yet

---

[6] "The Undiscovered Self," *Civilization in Transition,* CW 10, par. 547.

with no memory of traumatic experiences. Sometimes they can recall stressful or debilitating environments that persisted throughout later childhood, from which they extrapolate backward to imagine what it must have been like in early childhood. But actual experiences of trauma usually elude them. Thus, this book. How can we recognize early childhood trauma when there are no memories of traumatic events? How can we discern the adaptations that the archetypal dynamics of the unconscious put into place before consciousness developed? And, finally, can we rework the past? Are we condemned to live with such sequelae forever or can the course of life be altered?

That is what we shall attempt to answer. We'll do it from a Jungian perspective, for that is the only psychology that both deeply and extensively explores the strange world of the unconscious and accords it a constructive role in psychological development. With that in mind, let's begin.

Psychoid

Psyche and brain are inextricably intertwined. Yet, we know very little about how they interact. Most of what we presume to know comes by way of analogy. Perhaps most useful are analogies with the universe. After all, we are born of the universe and participate in it throughout our lives. Furthermore, the basic dynamics of planets, stars, solar systems, galaxies, and clusters bear striking resemblance with what we now know about brain mechanics. So let's begin with an astronomical analogy.

Black holes comprise a galaxy's core. There, mass is so dense that its gravitational force allows nothing to exit, not even light. The line of demarkation that separates the realm of greatest mass and gravitational pull from that which is weaker is called the *event horizon*. We see only that which lies on our side of that line. Thus, the line represents a horizon beyond which we cannot see what is happening.

What would be the event horizon of the mind? How about the line between conscious and unconscious? Jung wrote that the collective unconscious "cannot be directly perceived or 'represented.' "[7]

---

[7] "Synchronicity: An Acausal Connecting Principle," *The Structure And Dynamics of the Psyche,* CW 8, par. 840.

The personal unconscious (the individual shadow) is equally shrouded in mystery. Dreams, projections, transference and countertransference reveal only a fraction of its dynamics and contents. As the collective unconscious is the domain of the physical chemistry that gives rise to psyche, Jung identified it as the realm of the "psychoid."[8]

We'll focus on that realm with an eye toward illuminating the mechanics of the psyche through recent scientific discoveries. Then, we'll apply that information to the individuation process. Our goal is to facilitate individuation in the consulting room and beyond.

Where It Begins

Chemical reactions are the domain in which psychology and physiology meet. In that domain, each influences the other. Actually, each is a function of the other.

That's pretty familiar. Take a drug, and our perceptions and behaviors change. Become excited, and our bodies change. Worry and fret too much, and we get ulcers. Catch a cold and we lose our appetite, want to sleep, withdraw from others.

The relationship between body and mind is so well established that we usually take it for granted. Yet, when we look beyond the obvious, when we analyze the mechanics of body and mind, things become very complex. The pathways through which physical and mental phenomena are generated, develop, proceed, transform, and discontinue become more complicated the more closely we look. And the more closely we look, the more we discover that the most powerful and complex activities of body and mind depend on the smallest of dynamics occurring at the molecular level. The greatest depends on the smallest.

Analytical Psychologists—Jungians, if you will—and all who join them in their research have been fascinated by this relationship for more than a hundred years. Moreover, Jung discovered that alchemists were exploring it centuries ago. Today, biochemistry and microbiology are describing its basic mechanics. More than ever, distinctions between mind and body are losing their usefulness.

---

[8] Ibid.

"Archetypal dynamics" is the term Jung gave to physical and psychological development and functioning. He postulated that those dynamics originate in tensions of opposites. Those, however, are not static, like a tug of war where both sides are in a stalemate. Rather, they oscillate. Back and forth they go, elaborating themselves along the way. Eventually, multiple tensions of opposites that carry the original patterns oscillate with each other.

Sound confusing? It should, for this is the realm of deterministic chaos. As I explained in *Archetypes*, deterministic chaos is where Analytical Psychology and chaos theory meet.

Archetypes are strange (*fractal*) attractors. To explore that world is to find ourselves in "the realm of the incomprehensible."[9]

Let's look at how Analytical Psychology and chaos theory apply to everyday life. We'll pay particular attention to how archetypal/chaotic dynamics point the way to development, whether prospective or retrospective. We'll see how the more problematic aspects of our lives may actually be in the service of transformation, either from earlier, less adaptive perceptions and behaviors, or from the inevitable backwaters of life that entrap us just when we think we've figured everything out.

We'll begin with a broad overview of how body and mind interact and then focus on how their interactions lead to development. In short, we'll look closely at *morphogenesis.*

Morphogenesis

Here's a simple question: Who is reading this? Most of us would answer: "Why, I am, of course!" But who are we? Are we the singular identities that we see ourselves as being? Are we the roles we play? Are we the biochemical reactions that make it possible for us to be or do anything at all (like read this book)? Just who or what are we? The simple question leads to numerous and complex answers.

We could increase the complexity by asking what life is all about. Why does what happens to us happen to us? What leads us to fall in love, to become excited by mystery, to spend endless hours, days, and years in

[9] From the epigram to "Chaos," the second chapter of *Archetypes* (p. 40).

the Academy? What ever possesses us to become Jungian analysts? Or anything else, for that matter?

While we're at it, let's throw some other questions into the pot. What goes on in the consulting room? Transformation? Healing? Cognitive restructuring? Active Imagination? Treatment? Endless illusions? Nothing? Why do we go into analysis in the first place? What drives us to play around with the forces at work in our bodies and minds?

Perhaps we are the unwitting agents of internal dynamics. Freud said we are enactors of drives, that we are driven to do what we do by forces within us that are only barely within our control. Jung said that our perceptions and behaviors are guided by—if not generated by—archetypes. But what are they? Where do they come from? And just what are their intentions?

We can answer that. Sort of. Archetypes are in the service of morphogenesis. In other words, archetypes are one means by which morphogenesis furthers our development. What, then, is morphogenesis?

The word morphogenesis comes from the Greek *morpho*, meaning shape, and *genesis*, creation. It refers to the process by which we are constantly created, altered, renewed, or edited. In other words:

> All embryos, from worms to humans, are shaped during development by morphogenetic steps that tug, bend, fold, and sculpt epithelial sheets into forms that resemble, or are the precursors of, the final adult structures.[10]

How this all takes place is the question that untold hours of research currently seek to answer. Morphogenesis appears to begin at conception and end at death.

Consider: In the beginning, a sperm and an egg combine and exchange their genetic material. That simple pairing of two cells leads to eyes, arms, hearts, nervous systems, and everything else that becomes a person when first it takes a breath. After that, the generation and editing of shape continue throughout life.

Biochemistry permeates all of it, even that which we ascribe to consciousness, for example moral, intellectual, artistic, and other forms of development.

---

[10] W. Razell and P. Martin, *Embryonic Clutch Control,* p. 1181.

It can all seem hopelessly complex. While it is certainly complex, it is not hopelessly so. There is much we can learn about the interactions between physical and mental processes. Just as physiology influences psychology, so too does psychology influence physiology. And therein lies the focus of this book. We'll look at physiology and psychology in terms of the relationships between the two. We'll find that chaos theory has much to say about those relationships. We'll see how the insights gained from chaos theory operate in the consulting room, between analyst and patient. Finally, we'll propose that relationship is the most archetypal of all archetypal dynamics, an understanding of which can improve our relationships with ourselves and with one other.

Parasites

Parasites are remarkable examples of how the body influences the mind. For example, have you heard of *Dicrocoelium dendriticum*? Probably not. I hadn't until I read about it in *Parasite Rex.*[11]

This little lancet fluke can procreate only in a grazing animal's gut—for example, a sheep's. Unfortunately, that which is in the gut ends up being excreted. Which *then* creates a rather interesting challenge: How can it procreate in the sheep's gut when it is no longer there?

Perhaps it could attach itself to the sheep's intestinal wall. No luck. Were it able to do so, the sheep's immune system would most likely make short work of it. So the problem is not so much how to stay in the sheep's gut, than how can the fluke get back into the sheep's gut once it has been excreted. The solution is so ingenious that consciousness (were this little helminth to possess that capacity) would be challenged to come up with it. Here's what it does: The parasite changes the sheep's feces. Through a complex biochemical dynamic it makes sheep feces very attractive to slugs (were we in need of a reason not to eat slugs, that would serve very nicely). The slug comes along, munches on the feces, whereupon the parasite ends up in the slug. Nice work.

Now the parasite changes the internal chemistry of the slug to make its slime very attractive to ants. Happy now to be excreted in the slug slime,

---

[11] Carl Zimmer, *Parasite Rex: Inside the bizarre world of nature's most dangerous creatures.*

it awaits the ants' arrival. Arrive they do, eat the slime, and our little buddy winds up in the ants. All well and good. But what does that accomplish?

Well, it's another step along the way back into the sheep. Now the parasite changes the body chemistry of the ants. Ants like to remain close to the ground. Any of us who lazed in the grass as children looking at those amazing microclimates under the grassy canopy remember seeing ants scurrying furiously to and fro on the ground under the grass.

Ants who have the parasite in them behave differently. Due to the changes in their body chemistry, they no longer prefer the ground. Now they seek the very tip-tops of the grass. They become heliotropic (or phototropic, as the case may be). Lazing in the sun on the tips of the grass, they suddenly find themselves ingested by grazing sheep. *Et voila!* Now the parasite is back in the sheep's gut, where it can procreate and begin the cycle all over again. Too bad for the ants. Great for the parasite.[12]

Another parasite reproduces only in a heron's gut. Again, this creates the problem of how to gain entrance to a heron once the parasite has been excreted. Like the *Dicrocoelium dendriticum*, nature has evolved a plan. It makes the heron's feces very attractive to fish, which eat the parasite along with the feces.

Now the parasite is in the fish, where it gains access to the fish's biochemistry. The fish, its biochemistry changed by the parasite, delights in flopping around on the surface of the water, where herons are attracted to its motion. The heron eats the fish, *et encore voila!* Parasite back in bird.[13]

What makes this all particularly interesting (apart from the complexity of its brilliance) is how parasites change the behavior of their hosts. Like the question of just who is reading this book, we may ask just what is changing the behavior of its host. Of course, like the answer "I am" to the question of reading, we would probably say "the parasite." But do parasites think? Do they analyze situations, recognize what is needed, discuss with their colleagues, and choose the means to implement the

---

[12] Ibid., pp. 87f.

[13] Ibid., pp. 107f.

changes that are necessary? Well, yes and no. Many parasites clearly change the behavior of their hosts to their advantage, which surely must involve some recognition of what is needed. But who or what is doing the recognizing and making the decisions?

The answer is relatively simple—molecules. As molecules encounter one another, they happily combine to form more complex molecules. They also divide other molecules into component parts, which may or may not reassemble into other configurations. Amazingly, it's all automatic.

Or is it? Can molecules choose how they change other molecules or what other molecules they change? Or is it all just an automatic chemical process? If this seemingly reflex, automatic behavior leads to results that seem deliberate and brilliant, is it all simply a matter of chance?

That's the question of the moment: How do supposedly spontaneous, non-conscious biochemical dynamics generate the reflections, analyses, and choices that are the hallmarks of consciousness? Or, in psychological language, how does the unconscious generate consciousness?

Genetics

Jung proposed an important technique through which to engage the unconscious in the service of greater consciousness, about which we'll have much to say in later chapters. For now, let's look at a current scientific explanation. While in Jung's day we knew very little about the intricacies of physical chemistry, we've learned a lot since then. Let's look at microbiology.

Most of us know something about genetics. We know that genes provide the blueprints for everything in our bodies. And we know that genes are composed of DNA. We may even know that when DNA is transcribed into RNA, the RNA can be translated into proteins, which fold in ways that produce all the substances and dynamics of the body.

If that seems complex, consider this: The DNA in each cell of our bodies is about six feet in length. Were we to cram all of the DNA in all of the trillions of cells in our bodies into a cube, it would be about the size of a typical ice cube. But if we laid all the DNA end to end, it would stretch to the sun and back between 70 and 400 times. Experts disagree on exactly how many times, but that's the typical range.

Inconceivable. But true! Does all that DNA constitute an immutable blueprint? Certainly a blueprint. But not immutable. Complicating the picture are transposons and epigenetics. Transposons are sections of DNA that wander around the genome looking for places to insert themselves, whereupon they alter the alleles (genes) that determine everything else. Plasmids are larger segments that move greater distances within the genome. They also routinely form themselves into circles. While few Jungians could fail to see the symbolism in that, we'll leave it for others to pursue.

There is one form of transposon that we all recognize: Viruses. Unlike the smaller transposons that move only within their own genome, viruses can move from one genome to another. These bits of genetic material are not officially alive. They cannot reproduce and have no mitochondria, the energy-producing furnaces in cells that make them self-sustaining. Viruses have more in common with rocks than with anything living.

Yet what havoc viruses wreak! They permeate cell membranes and hijack their reproductive machinery in order to replicate. Once accomplished, they burst out of the cell and infect others. Cells lend their reproductive capacity. We provide the means of transmission: we sneeze, cough, shake hands. In that capacity we are little more than mass transit systems.

Viruses are very much like parasites. They hijack their hosts to assist in their replication and spread. And like the unconscious, they seem to manifest agency, intent, and design.[14] But they are not alive. Just what is doing all of that? How can something that is not alive act like an agent, actualize intent, and implement design? Were the alchemists correct? Is there an animating force in matter? If so, is it morphogenesis?

Actually, it is. *Mutations* refers to the innumerable ways in which the chemical composition of DNA can be changed. But there's more. There is also epigenetics (*epi* = beyond; so *epigenetics* = beyond genetics).

---

[14] See my elaboration of this idea in "Switching Tracks: Parallel Paradigms in Psychology and Religion."

Epigenetics

Epigenetics refers to those processes that govern how DNA is transcribed into the RNA and RNA is translated into proteins. Epigenetics explains how those processes are implemented. Changes in epigenetic processes produce changes in outcome.

Enzymes are the primary agents of epigenetics. They make up the chemical soup in which transcription and translation occur. As the soup changes, so does the outcome of transcription and translation. We can think of genetics as construction materials and epigenetics as contractors. Enzymes influence how proteins fold. Remember: Proteins are the building blocks of everything.

But some proteins seem to misfold. These are called *prions*. Prions are generally accused of producing aberrations. While we may not recognize the condition called bovine spongiform encephalopathy (BSE), we may be familiar with its shorter name: mad cow disease. *Bovine* refers to the cow, *spongiform* to the appearance of the brain in mad cow disease—it looks like a sponge. Great gaps appear in the grey matter. Much is missing. The brain (*encephalo-*) is diseased (*pathy*).

Sounds pretty grim. Those little prions/misfolded proteins are really a menace. Or are they? In fact, they make it possible for genomes to develop in new ways. For example, yeast is able to adapt to new conditions thanks to prions, which create new versions of proteins that often improve the survival of the yeast. Think of prions as mutations that can enhance—or undermine—survival.

The dynamics of epigenesis are so complex that only now are we beginning to figure them out. The chemistry behind those dynamics, however, is even more complex. Complexity compounded by complexity. But as we saw in *Archetypes,* complexity is often based on some very simple processes, among which are those to which we refer as *chaotic*. Later on we shall have much to say about the role of deterministic chaos in the morphogenesis of individuation.

Antibodies

Transposons, plasmids, viruses, prions—science fiction come true. The stuff of genetics and epigenetics is positively astounding. That all of this—over which we have no more than a modicum of influence—goes on inside us every second of the day is either fascinating or frightening. Nevertheless, it is absolutely brilliant, yet with none of the characteristics of that to which we usually ascribe brilliance, namely, consciousness.

For our final example, we'll cite antibodies. These molecules help us to neutralize diseases in the present that we have had in the past. Antibodies are how our immune systems adapt to diseases. Thus, they are part of the adaptive immune system, which develops over time. This contrasts with the innate immune system, which is present at birth.

On the opposite page is a schematic drawing of an antibody and accompanying antigens. Antibodies are formed from a type of blood cell called a *B cell*. On the B cell's surface are receptors that become activated when they detect an antigen (a molecule that does not belong to our body, thus not "self"). B cell receptors then alter their chemical makeup until they find the right combination to bind with the antigen. To do this, they *hypermutate*, that is, alter their genetic code at a very high speed until the right combination emerges.

In the image, it is the V part of the Y that engages the antigen. There are four chemical domains on each branch of the V, each of which starts spinning like the dials on a slot machine until they hit the jackpot. Once it can engage the antigen, the other end—the I end of the Y—detaches the receptor from the B cell. The receptor is now an antibody, which sends out a chemical signal for the innate immune system to arrive on the scene and deconstruct the antigen. With that, the threat is gone.

Pretty impressive! Yet all the more so when we consider whether or not B cell receptors somehow know what they are doing, can analyze situations and adapt in a way that gets the job done. We have yet to identify what, if anything, comprises a center of consciousness in a B cell receptor. All we know is that the amazingly ingenious solution to the problem of infection is carried out by diverse molecules within the receptor. Apparently, it's all automatic.

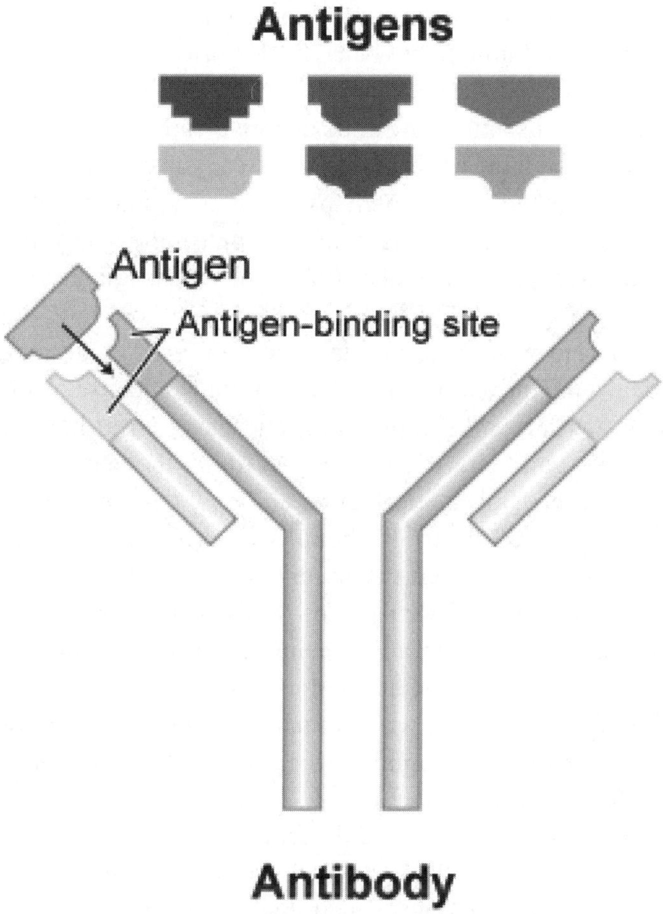

Like with parasites and transposons, there's really nobody home. This is physiology "behaving" in ways that look an awfully lot like consciousness—or at least awareness. But while it may process like consciousness, we can find no functioning center of that consciousness. No ego. Or, at least, that's the way it appears.

Even more intriguing, the complexity of the physiological processes—their awareness, ability to adapt, range of solutions, level of acumen, high efficiency, and yes, speed!—often seem much more sophisticated than consciousness. Like the unconscious, which is better equipped than consciousness to handle lightning-fast decisions,[15] these processes can size up situations and implement behavior far faster and more effectively than consciousness.

Perhaps the unconscious is just this adaptive capacity of physiological processes. If physiological processes can recognize and deal with the countless challenges to health, perhaps the agency, intent and design of the unconscious is precisely this ability of physiology. In other words, could the unconscious be the discerning and acting quality of organic matter?

As you may have guessed, that is what this book is all about. From here we'll leave the question of whether or not the unconscious is a function of matter (as far as we know, it is) and move on to how that plays out in perception and behavior. In other words, if the unconscious is the name we give to physiology's adaptive abilities, then maybe we should try to listen more carefully to it. Should we not recognize when it is at work and learn how to supplement or edit its activities?

That, of course, is the whole point of Analytical Psychology. Working with the unconscious, as opposed to seeing it as just one more obstacle to the ego's designs, is what Jung advised us to do. The core of his writings are geared to that pursuit. He wrote much about how to perceive the unconscious at work and how to judge when to cooperate with it and when to resist it. Well, more about that later.

Now we'll elaborate an essential element in Jung's theory, namely how consciousness can collaborate with an unconscious that, however brilliant, often seems quite obtuse, as if there's nobody home.

The Ghost in Our Machines

Consider the following: A CEO dreams that an employee has put a dirty, greasy pan in a sink at the office where the CEO washes his face. What could that be about? In search of an answer we focused on the CEO's professional persona, which perhaps was being dirtied by this employee.

---

[15] Stepping off a sidewalk into the street and hearing a car horn, for example. See below, Chapter Four.

Sure enough, on closer inspection, the CEO discovered that the employee had made some seriously misguided financial decisions that could sabotage the corporation. Fortunately, the situation was corrected and disaster was averted.

Or how about a woman who consistently dreamed that she was being chased down a dark hallway by an unknown assailant bent on raping and murdering her. One night, when she found herself again in the dream, she mobilized a sufficiently slim vestige of consciousness to stop, turn around, and confront her pursuer. She was surprised to find that he was a six-year-old boy who was lost and could not find his way home.[16]

This can happen collectively as well. In January of 2011, I was awaiting permission to enter Gaza through the Rafah Crossing in Egypt. After three days I was allowed in (well, allowed out of Egypt. Then I had to obtain permission to enter Gaza, which took all of ten minutes). Only then did I learn that a man had set fire to himself in a market in Tunisia. My Palestinian friends reported that Tunisia was now in an uproar. People were demanding that their autocratic leader resign. Several days later, he did.

"Egypt is next," my friends told me. And sure enough, about two weeks later, people began assembling in Cairo and demanding that Mubarak resign. Soon, the Arab Spring was blossoming throughout the Middle East. When the challenge to arbitrary authority reached Wisconsin, a pizza parlor in Madison put a notice on its website that people could donate money for pizzas for the demonstrators. Donations streamed in, some from Egypt!

How does this happen? Maybe it's the internet, with its instant com munication. Certainly social networking sites can put people throughout the world in touch with each other. But that's not new. Something is moving in the collective psyche of humankind. The *anima mundi* (soul of the world) has apparently had enough of autocrats. As Emma Jung wrote, much depends on the times.[17] For authoritarian rule, that time may be now.

---

[16] We'll have more to say about this person in Chapters Six and Eight.

[17] See her *Animus and Anima,* p.7 " the law of the time-moment, the *kairos*, to which we are all subject and from which we cannot escape."

Morphogenesis Redux

To repeat: Human morphogenesis seems to begin at conception and end at death.[18] But just who or what is running the show before consciousness is up and running? Who or what determines our perceptions and behaviors before consciousness develops sufficiently to observe, reflect, analyze, and choose?

Two-week-old babies mimic the languages spoken around them. The cries of infants in France sound like French and in Germany, German.[19] Obviously, babies do not lie there thinking: "Oh, that didn't work. I think I'll try crying in a way that sounds more familiar to my parents." Babies also cry in ways that particularly penetrate the hearts of their parents. So, who or what decides on that?

Jung would say that the unconscious is responding to the environment in ways that determine the infant's behavior. He would add that the development of the infant, (including such mimicry, which is in the service of increasing the infant's potential for survival) is governed by archetypes.

So, what are archetypes? If they are the agents of development, how do they "know" what we need? How do they get us to follow their intent? Do archetypes possess agency, intent, and design? If so, how?

We don't know. We cannot open the brain and see what is going on in this respect. The technology that tries to make that possible shows little more than that which accompanies perception, thought, and behavior.[20] To see the *what* is not necessarily to discover the *why*.

Is science useful here? Can it help us with any of this?

Actually, yes. From the moment of conception cells multiply and diversify spontaneously and automatically. We have come to use the term

---

[18] Presumably. We don't know what—if anything—exists after death. Curiously, for a brief time after we die, our hair and fingernails continue to grow. And while the *I* with whom we identify may be unrecognizable, the molecules that participated in that identity continue on, combining and recombining in support of whatever morphogenesis requires.

[19] See Friederic Mampe and Christophe Wermke. "Newborns' Cry Melody is Shaped by Their Native Language," in *Current Biology,* vol. 19, no. 23 (November 2009).

[20] Some scientists say that brain imaging is little more than phrenology. See Ed Diener, and Russell A. Poldrack in the bibliography.

"self-organization" to describe these biological processes that are inherent in molecular dynamics and increase the complexity of the organism. However, note the words *inherent, dynamics,* and *complexity*. Each represents a phenomenon that we may be able to describe, but that we can hardly understand. The words are essentially euphemisms. Self-organization remains as elusive as before.

Nevertheless, biological processes are becoming increasingly accessible. For example, let's look at how sperm and egg generate a human being. The first cells to develop after conception are neural cells. The dynamics that give rise to those cells are controlled (initiated and sustained) by neural precurser cells (NPCs), which activate and regulate glial cells, which mylenate the nerves through oligodendrytic cells.

Something has to be in charge of that process. Is there an epigenetic factor? Here's the current explanation:

> Wnt signaling (through Wnt ligands) is necessary for the conversion of NPCs into glial cells. But that can occur only when histones and chromatin come into the right configuration (when there are "key intrinsic changes in histone H3 acetylation and trimethylation at the ngn1 and ngn2 promoters in NPCs, such that, at early stages…[there is] high acetylation and low methylation, but at later stages, low acetylation and high methylation).[21]

Pretty complex. Way too complex for our needs. But it does resonate with our understanding of archetypal dynamics. The first resonation is that, like Jung's language of psychological dynamics (*archetypes, complexes, unconscious, tensions of opposites, active imagination, networks of associations, canalization of energy, libido, ego*), microbiology also has its own language.

Language is the key to understanding anything. No words, no ability to communicate, share, discuss, reflect, define. Without words, we cannot even articulate what we perceive. Language creates meaning. In the above description of the conversion of NPCs into glial cells, a picture of what happens comes into focus in the words used. So, where do the words come from? They are simply assigned. While *Wnt signaling* may seem like an arbitrary phrase, it derives from observations and analyses

[21] Marc R. Freeman, "Specification and Morphogenesis of Astrocytes," p. 774.

of the dynamics for which it is named. The words assigned to the dynamic essentially become the dynamic.

The second resonation is the bifurcation and oscillation (tensions of opposites, in Jung's language) on which the dynamic is based. The words "...at early stages…[there is] high acetylation and low methylation, but at later stages, low acetylation and high methylation" describe how genes are turned on and off to generate the dynamic in question. Acetylation is the process by which genes are turned on. Methylation turns them off.

In other words, in this brief account of how fetal development proceeds in its earliest stages, we can see how the scientific description of morphogenesis corresponds with Jung's. Both require the creation of a language to describe phenomena. Furthermore, the basic dynamics at work are the same for both: Oscillating tensions of opposites. As we shall see in Chapter Three, those basic dynamics are consistent with chaos theory.

In *Archetypes*, I explained how chaos theory corresponds with Jung's metapsychology. Science finds that chaotic dynamics are at the heart of most biological processes, including how the brain works.[22] Could this be the link between Analytical Psychology and brain science?

Absolutely. The dynamics that Jung described as the foundation of psychological development have their analogs in science today. That much of neuroscience (as well as genetics and immunology) parallels Jung's metapsychology gives us license to consider how science can expand our understanding of how archetypes function in individuation.

For example, schizophrenia is considered to be a function of deterioration of communication among neurons.[23] Early in his career Jung recognized that restoring relations between ego and unconscious was highly therapeutic and could reverse the onset of mental illness (especially neuroses). In today's language, we could say that imaginal techniques like active imagination increase synaptic complexity and integrate functionally autonomous brain regions.

Since relationship is at the heart of all the dynamics we are exploring,

---

[22] We'll have more to say about that in the section on synaptic plasticity.

[23] Particularly in the hippocampus. Dopamine regulation is also highly implicated.

we can say that relationship is an archetypal dynamic. Might relationship be the fundamental dynamic through which archetypes operate? Is it the archetypal dynamic of archetypes?

Probably. Relational dynamics among the smallest components of the human organism have enormous effects. A bacterium that enters our bodies and consorts with our smallest molecules can lead to our death. The complex dynamic that constitutes an organism may be undone by the dynamics of its smallest components. The integrity of the whole depends on the behavior of its parts. It all depends on which relationships win out.

Childhood Trauma

What does all that have to do with childhood trauma? Traumatic experiences and environments disrupt relationships among the functioning parts of a child's developing psyche. The unconscious is forced to assemble ways of dealing with those disruptions. That, however, can disrupt future relations between conscious and unconscious if the former is consistently preempted by the latter.

When the sequelae of childhood trauma appear in the consulting room, healing depends on adjusting the balance between conscious and unconscious. That is accomplished through relationship—between analyst and patient, between conscious and unconscious. As we shall see, by engaging problematic relationships between conscious and unconscious that originated in childhood, the analytic relationship reworks the past. Reworking the past heals those relationships.

In other words, relationship heals relationship by activating aspects of relationship previously suppressed due to traumatizing environments. Jung gave us techniques by which to accomplish the task. His explanations of how those techniques succeed is grounded in late nineteenth- and early twentieth-century thought, however, leading them sometimes to seem archaic in light of the science of our day.

To remedy the situation, we'll draw upon contemporary science to illuminate some of the more abstruse ideas in Jung's metapsychology. For example, imaging techniques and biochemical analyses that were unavailable in Jung's time greatly expand our understanding of the physical chemistry underlying psychological processes.

From the Jungian perspective, archetypes lie at the heart of those processes. So in the next chapter we'll look at the role archetypes play, how they invest psychological processes with meaning, purpose and goals. In subsequent chapters we'll look at the patterns that archetypal dynamics generate. Then, in Part Two, we'll elaborate how to apply these ideas in the consulting room, particularly how to deal with adaptations that were put into place in childhood, before consciousness was up and running.

We've looked at how physiology influences psychology. Now let's look at how Jung described that relationship. Let's take a closer look at archetypes.

2
Archetypes: The Dark Matter of the Psyche

Archetypes

The observable universe contains only a fraction of the mass of the total universe. The rest is called *dark matter*. Dark matter is like black holes. We can measure its influence, but we cannot see it. The psyche is similar. We can see what it does (mainly in perceptions and behaviors) and even manipulate it. But we are far short of seeing its component parts. As noted earlier, even digital imaging techniques may amount to little more (at this point) than phrenology.

Jung designated archetypes as the link between psyche and soma, mind and body, thought and biochemistry. He believed that the psyche reveals itself in the images that it produces, particularly dreams, folktales and myths. In the case of the latter two, he noted that individualized personal elements had been filtered out through collective reflection over time. Thus, dreams, folktales and myths are like case studies, which reveal the psychological dynamics of the human psyche.

That images comprise an essential aspect of consciousness is now being reinforced by cognitive science. For example:

> I propose that perceptual analysis results in redescriptions of spatial structure in the form of image-schemas. These redescriptions constitute the meanings that infants use to create concepts of objects, such as animate and inanimate things, and relational concepts, such as containment and support. I further propose that image-schemas provide a level of representation intermediate between perception and language that facilitates the process of language acquisition.[24]

If correct, Dr. Mandler confirms that images (image-schemas) represent in apprehensible form the complexity and nebulosity of the psyche's response to everyday experience. They are also essential catalysts for the conversion of perceptions into language.

That corresponds with Jung's description of how information from

---

[24] Jean M. Mandler, "How to Build a Baby: II. Conceptual Primitives," in *Psychological Review,* vol. 99, no. 4 (October 1992), pp. 587ff.

everyday life is converted into psychological insight. *Archetype* is the term that Jung assigned to that in the psyche which invests certain images (or image-schemas) with a compelling quality that promotes psychological development. Archetypes invest images with a valence that either attracts us to or repels us from the life experience represented by that image.[25] They are navigational aids on the trackless ocean of life experience, so to speak.

Here's what Jung had to say:

> The archetypes are, as it were, the hidden foundations of the conscious mind, or, to use another comparison, the roots which the psyche has sunk not only in the earth in the narrower sense but in the world in general. Archetypes are systems of readiness for action, and at the same time images and emotions. They are inherited with the brain structure—indeed, they are its psychic aspect. They represent, on the one hand, a very strong instinctive conservatism, while on the other hand they are the most effective means conceivable of instinctive adaptation. They are thus, essentially, the chthonic portion of the psyche, if we may use such an expression—that portion through which the psyche is attached to nature, or in which its link with the earth and the world appears at its most tangible. The psychic influence of the earth and its laws is seen most clearly in these primordial images.[26]

"Archetypes are systems of readiness for action" tells us what archetypes are. That they are "inherited with the brain structure" tells us where they originate. Interestingly, where they originate tells us a lot about what they are. That they comprise "that portion through which the psyche is attached to nature, or in which its link with the earth and the world appears at its most tangible," means they are functions of physical chemistry. Thus, they are not *homunculi* residing in the psyche that behave like little consciousnesses. Rather, they are automatic molecular reactions to chemical events that have evolved over eons of human history to attain an adaptability that rivals—and often exceeds—that of consciousness. As such, they are the agents of morphogenesis in the psyche. Or, perhaps more specifically, they are the term we give to the dynamics that

---

[25] See my *Archetypes,* pp. 89ff.

[26] "Mind and Earth," *Civilization in Transition,* CW 10, par. 53.

give rise to morphogenesis. To connect morphogenesis with molecular dynamics is to explore the physical basis for archetypal dynamics.

Creating a dialogue between Jung's ideas and contemporary science expands our understanding of the mechanics of psychological development. The insights gained improve our ability to perceive, understand, and cooperate with those developmental impulses, however shrouded in unconscious confusion they may be. In Part Two, we'll apply that understanding to clinical case material.

Archetypal Dynamics

"Archetypal dynamics" is the term Jung gave to the psyche's ability to perceive and to engage the environment outside the realm of consciousness. Before consciousness is up to the task, there are only archetypal dynamics. Once consciousness develops the ability to perceive, decide, and implement adaptation and growth, it does so alongside the unconscious's already up-and-running efforts.[27] This can lead to competition, conflict, and even sabotage of the one with or by the other. Ideally, conscious and unconscious can be brought into a collaborative relationship that furthers morphogenesis.

Archetypal morphogenesis can be either individual or collective. That is, transformative experiences can happen to both individuals and groups. Individual experiences happen every day. Group experiences are a bit less frequent. Events throughout the Arab world in 2010-11, the declaration of the Thirteen Colonies as a United States in 1776, even the mass suicide at Jonestown (in 1978) all qualify as group transformations.

That transformation can be positive or negative reflects the bipolarity of archetypes. Jung stated that archetypes can seek either the spiritual or the chthonic pole. Viewed from the perspective of organisms as a whole, *spiritual* = integration into greater complexity, and *chthonic* = disintegration into lesser complexity. As I explained in *Archetypes*, the former involves deterministic chaos and the latter entropic chaos.[28] We'll have much to say about that in the next chapter.

---

[27] *Unconscious* is the term we give to everything of the psyche that is not conscious.

[28] *Archetypes,* pp. 45f, 133ff, 143ff.

In describing the influence of archetypes, Jung made a clear distinction between teleology and finality. He wrote that archetypes do not seek particular goals (teleology). Rather, they move development in particular directions (finality). That has great significance for the role of the ego, which must avoid the delusion that it knows exactly what archetypes are "trying" to accomplish. As conscious beings, we can determine only their direction. Our role is to accompany, cooperate with, or modify their efforts.[29]

Archetypes generate tensions of opposites. This is the first stage in consciousness-raising, for that is when differentiation becomes possible. Without differentiation, nothing can be distinguished from its context. All is just an amorphous blob.

When an archetype becomes active, however, it establishes a tension between what is and what needs to be, between being and becoming. That tension, however, is not static, like a tug-of-war between two equals. Rather, it is dynamic. There is an ongoing back and forth between being and becoming as each supercedes the other.

As dynamic entities, archetypes are pure presentation and cannot be represented. They are visible only through the stuff they activate. As Jung said, archetypes are pure forms that become visible only through their influence on contents. Contents are the stuff of images. Archetypes are the compelling quality the image manifests. Archetype + image = compelling image, which Jung defined as a *symbol*.

That makes archetypes powerful originators of morphogenesis in the psyche. Each archetype is "a dynamism which makes itself felt in the numinosity and fascinating power of the archetypal image."[30] Archetypes are not things. They are dynamics. It's like a novelist, who is "not a person, nor even a personality but a process that has resulted in a sequence of texts."[31]

---

[29] This idea is reflected in contemporary theories of organizational development. See, for example, Dietrich Dorner, *The Logic of Failure.*

[30] "On The Nature of the Psyche," *The Structure and Dynamics of The Psyche,* CW 8, par. 414.

[31] Joyce Carol Oates, quoted by Ann Hulbert in her review of Oates' *A Widow's Story,* in *The New York Times Book Review* (February 20, 2011), p. 10.

Furthermore, research into brain dynamics has found that viewing optical illusions precipitates

> sudden simultaneous activity in several parts of the brain, specifically the frontal, parietal, and temporal cortices of both hemispheres. [Consequently], …conscious perception is associated with a sudden global reverberation of neuronal activity, about 300 ms [milliseconds] after the stimulus.[32]

Moreover,

> the mental activities surrounding consciousness, perception, and thought are the result of columns of neurons stacked next to each other, forming elemental structures that function as a unit, with multiple assemblies of such units within the brain working together as a network of distributed, sometimes recursively connected, interacting brain regions with the different areas making specific and often task-modulated contributions.[33]

These two quotations make amply clear that brain activity is a relational endeavor. That is, components of the brain work together to produce (among other things) the phenomenon we refer to as *consciousness*. If relationship among elements in the unconscious (*simultaneous activity; function as a unit; working together as a network; connected, interacting brain regions,* in the above quotations) is the *sine qua non* of brain activity, then relationship among the various aspects of the psyche is the basis for psychological development (in Jung's parlance, *individuation*).

In other words, the morphogenesis of the psyche depends on relationship. At the end of the last chapter we stated that schizophrenia is a function of deterioration of communication among neurons. Specifically,

> multiple brain regions have been implicated in schizophrenic pathology including the hippocampus, subdivisions of the prefrontal cortex, the thalamus, and the striatum. Most likely, it is the *disrupted communications* between

---

[32] *Science,* vol. 318 (October 19, 2007), p. 361.

[33] John T. Cacioppo and Jean Decety, "What are the Brain Mechanisms on which Psychological Processes are Based?", in *Perspectives on Psychological Science,* vol. 4, no. 1 (January 2009).

these regions that leads to the manifestation of schizophrenic symptoms.[34]

If archetypes are the architects of morphogenesis in the psyche, they can operate effectively only if there is communication among the various aspects of the psyche. From the Jungian standpoint, the relationship between conscious and unconscious is critical to that communication. As with schizophrenia, disruptions of that relationship lead to what we call *mental illness*.

When the Ghost in the Machine Becomes a Demon

When all goes well in individuation, we rarely concern ourselves with what's going on in our minds. When things run smoothly, there are no wrinkles to interrupt our reverie. When life does not go well, however, we sit up and take notice. We want to know what's going on, why it's not going right, why we feel bad.

First we try to discover what it is. We label it with diagnoses. Then we try to correct the situation. Unfortunately, as Dostoevsky wrote: "I called on the entire medical profession: they diagnose beautifully, they tell you what's wrong with you one-two-three, but they can't cure you."[35] How often do we feel that? Nevertheless, there are ways to heal the psyche when it has been displaced from its optimum functioning.

Individuation stalls when communication and relationship are disrupted. Given that the brain contains upwards of 100 billion neurons that communicate with each other through upwards of 100 trillion synapses, opportunities for disruptions are astronomical. Fortunately, Jung's structure and dynamics of the psyche lump these enormous numbers into two categories: conscious and unconscious. Our concern is the disruption in communications between the two. Healing those breaks in relationship then becomes our primary challenge.

What is the relationship between archetypes and mental illness? First, let's clarify what we mean by mental illness. Harry Stack Sullivan wrote:

---

[34] S. Hemby and J. O'Connor, "Transcriptional Regulation in Schizophrenia," in D. Romer and E. Walker, eds., *Adolescent Psychopathology and the Developing Brain: Integrating Brain and Prevention science*, pp. 105-6; my italics.

[35] Fyodor Dostoevsky, *The Brothers Karamazov,* p. 641.

> I believe that there are no peculiarities shown by the morbid; there are only differences in degree—that is, in intensity and timing—of that which is shown by everyone.[36]

In other words, this "Father of American Psychiatry" saw mental illness is a matter of degree, not kind. We are all to some degree psychotic, anxious, depressed, traumatized. Only when those states reach a certain level do we define them as mental illness.

Interestingly, that level differs from culture to culture. Psychosis is a good example. Behaviors considered normal in some cultures are in others labeled as mental illness (or *morbid,* as Sullivan says). Likewise, depression is more acceptable in some cultures than in others. Even certain contexts within a culture allow for greater extremes than in others. At sporting events, for example, it is often acceptable to act in ways that otherwise might be considered disordered. Sullivan also wrote that:

> Whenever I speak of dynamisms I am discussing universal human equipment, sometimes represented almost entirely in dreadful distortions of living, but still universal. And the distortions arise from misfortunes in development, restrictions of opportunity, and the like.[37]

In Jungian terms, "universal human equipment" would be defined as archetypes. As we shall soon see, "dreadful distortions of living" then refer to the activity of archetypes following "misfortunes in development", "restrictions of opportunity, and the like". For our purposes, then, mental illness is the result of archetypal adaptations that later on perseverate as maladaptive after-shocks in the system.

How do archetypal dynamics become maladaptive? It's actually pretty simple. Archetypes direct our attention toward that which furthers development and away from that which retards it. But they can focus our attention only on what exists. For example, environments differ in their capacity to provide mothering. Available resources for archetypal valencing can vary widely.[38]

---

[36] Harry Stack Sullivan, *The Interpersonal Theory of Psychiatry,* p. 305.

[37] Ibid.

[38] *Valencing* refers to the positive or negative charge that archetypes assign to aspects of our environments.

If mothering is required at a certain stage of development, archetypes can attract us to or repel us from only what is available. If the adults in our home are alcoholics, drug addicts or mentally ill, we may find ourselves attracted either toward the inadequate adults or toward some amorphously existential possibility that exists only in fantasy.[39] Either way our development is compromised.

The result is a split in archetypal awareness. While there is a press toward connecting, there is a corresponding awareness of the liabilities involved. We become split. We need to connect, but that to which we are connecting is toxic. That leads to a chaotic dynamic whose bifurcations are characterized by ambivalence and whose attractors oscillate between disappointment and longing, fear and the need for security, betrayal and the inclination to trust, anger and affection, and so on.

Remember: The "we" who experience this in the first several days, weeks, months and years of life are not the "we" who observe, analyze and decide later in life. Early in life, the unconscious is in charge. By the time the ego is ready to assume command, the deterministic chaos of early childhood has resulted in patterns that can frustrate, if not defeat, the ego's role. Thus Sullivan's words above are particularly apt.

Symbols

As we shall see, the only way to free the psyche from the compounded effects of deterministic chaos is to create a new bifurcation. That, however, is more difficult than it may seem, for the psyche expresses itself through symbols, which by their very nature can be annoyingly inscrutable, which makes them particularly vulnerable to literal interpretation.

Jung defined a symbol as the infusion of an archetypal form with the stuff of everyday life. Every archetype is "a vessel which we can never empty, and never fill. They have potential existence only and when they take shape in matter [they] are no longer what they were."[40] Thus, every

---

[39] Actually, since all environments are to some degree inadequate, such fantasies are part of childhood experience. We also encounter them in folktales and myths.

[40] "The Psychology of the Child Archetype," *The Archetypes and the Collective Unconscious,* CW 9i, par. 301.

symbol is "the best possible expression for an unconscious content whose nature can only be guessed, because it is still unknown."[41]

Moreover, "the symbol is a pointer to the onward course of life, beckoning the libido towards a still distant goal."[42] When an archetype directs our attention toward that which furthers individuation and away from that which doesn't, it does so by valencing something in the environment. Anything so valenced automatically becomes a symbol. In fact, "the psychological mechanism that transforms energy is the symbol."[43] It influences how we mobilize our efforts.

Nevertheless, we must always bear in mind that the unconscious is pretty crazy. Generated by the most powerful dynamics of the brain, namely, synaptic biochemistry, it often expresses itself in exaggerated form.[44] If my individuation requires me to confront my boss about his abusiveness, my dream the night before might be of walking into his office and blowing him away with an AK-47.

We must beware of taking either the content or the power of symbolic images literally. For example, I recently met with a man in the Middle East who was afraid of ghosts *(jinns).* He had previously been coming home from school at night when some militants who were patrolling the neighborhood shouted to him from the shadows: "What's your name?" Since then, he was terrified of going out alone in public.

Where do the ghosts come in? If they are symbols produced by an archetype, what is the contribution to individuation?

The answer appears when we consider the emotions of the militants. Patrolling the neighborhood occurs in an atmosphere of threat. At any moment, they could be attacked. When they confronted the young man, he heard two things. Firstly, he heard their fierce aggression and bold challenge. Of that, he was aware. Secondly, their bravado was permeated by wariness generated by fear. Of that, he was not aware.

---

41 "Archetypes of the Collective Unconscious," ibid., par. 7n.

42 "Schiller's Ideas on the Type Problem," *Psychological Types,* CW 6, par. 202.

43 "On Psychic Energy," *The Structure and Dynamics of the Psyche,* CW 8, par. 88.

44 Again we come to the paradox of the smallest molecules having the biggest impacts.

But his unconscious was. The unconscious is particularly astute at recognizing what consciousness often misses. Consider, for example, the friendly used-car salesman who asks: "What would it take to put you in this car?"

Ostensibly, he's just trying to give us the best deal. Our unconscious, however, immediately picks up on what he is not expressing, namely that the car is a piece of junk. We feel a strange presence of something that contradicts what we are experiencing.

Although the militants were suppressing any indication of their own vulnerability, the young man's unconscious picked up on it. But he didn't realize that. So, his unconscious presented it to him in the symbol of the ghost, which is something that is present but cannot be seen (at least not by objective observers). The ghost that he feared reflected the unconscious's attempt to provide insight into the situation. The fear it elicited in him replicated the fear the unconscious recognized in the militants.

Childhood Synaptic Adaptations

That phenomenon—the unconscious recognizing elements of life situations that the individual misses—begins at birth. Remember that babies' cries mimic the language spoken around them as early as two weeks after being born. Remember our question of just who or what was doing the recognizing and mimicking? Now we're getting close to the main point of this book.

As aspects of the mind, archetypes leave their traces in brain chemistry. Every life experience leaves its imprint on the brain in the form of synaptic configurations. Learn to play the piano and synapses form and develop around how our eyes and fingers interact. If we stop playing the piano for twenty years and then sit down to play again, those configurations (albeit a bit diluted) make it possible for us to begin playing immediately. We call it *learning*.

Learning how to play the piano is done consciously. But synapses form even when we are not aware that we are learning. And we are learning from the first day of our lives. Consequently, synapses begin configuring before we are conscious enough to know it. By the time we are aware, synapses have been configuring themselves for years. They are up and running. The question is the degree to which we are in tune with it—or not.

The ways in which our brains have learned to function—to perceive, analyze and cope with life events—persist after we decide to "take the helm" ourselves. If we are in accord, all goes well. If not, we experience the wrinkles in the fabric of life mentioned above. At that point, we have to adjust, either by cooperating with the unconscious or—as we often tend to prefer—by suppressing it.

Therein lies the origin of the most intractable of tensions between who we are and who we believe we ought to be. As with schizophrenia, the problem lies in lack of relationship, in this case between conscious and unconscious. Fortunately, Jung has given us a language and a conceptual framework not only to recognize our dilemmas but to solve them as well.

Our challenge is to engage the synaptic configurations that were encoded before consciousness was sufficiently developed to cope with the challenges of life. The unconscious is not some existential dump into which everything that evades consciousness winds up. Rather, it is an active organ of development and adaptation, of morphogenesis. Archetypes are active from the moment of birth. Only later does consciousness develop sufficiently to become their rival.

Archetypes and Morphogenesis

From here on in Part One we'll develop in detail the underlying dynamics of psychological development with regard to deterministic chaos, the paradox of defenses, and trauma. Then in Part Two we'll look at the life events that can throw it off course and what we can do about it.

We'll find that everything depends on relationship. Disturbances in individuation are caused by disturbances in relationship. Those must be healed in order for individuation to proceed. Our task later in life is to engage those aspects of our personalities that we had to leave behind in order to grow and survive. Back then, we feared that if we didn't abandon them we'd be forever handicapped by them.

Later in life, when we have developed sufficiently to go back and get them, they call to us through our emotions, perceptions and behaviors. Often, that feels like mental illness. But it isn't. It is simply that synaptic configurations formed when we were young are now firing off in our present circumstances. It is a call to action.

And act we shall. Just how is the focus of Part Two in this book. For now, we've seen that physiology influences psychology and how Jung conceived of the relationship between the two. Now, let's look at a brief review of chaos theory, which characterizes the psychodynamics of conscious and unconscious. Then we'll see how paradoxes unite chaos theory with life experience. Finally, we'll see how trauma epitomizes the kinds of life experiences that destabilize the relationship between conscious and unconscious.

We'll leave for Part Two how to deal with childhood synaptic configurations that give rise to challenges later in life. There, we'll elaborate how to meet those challenges by being for ourselves the adults we needed earlier in life, but who weren't there.

With that we can finally be united with ourselves. For now, let's look at how chaos theory might help us to understand the more unsettling aspects of childhood morphogenesis.

3
Chaos Theory

In this chapter we'll explore how the chaotic nature of psychological dynamics promotes healing and individuation. We'll pay particular attention to sensitive dependence on initial conditions (SDIC), self-similarity across scale, fractal dimension, the *r* factor, bifurcations, oscillations, iteration, and mutual inhibition. We'll keep it brief and as simple as possible. For a more complete elaboration of the relationship between chaos theory and psychodynamics, please see *Archetypes*.

Chaos Theory

Chaos theory distinguishes between two types of chaos: *entropic* and *deterministic*. Entropic chaos devolves into disorder and never returns to order. Deterministic chaos oscillates between order and disorder while generating patterns that appear and reappear as the dynamic *iterates* (from Latin *iter,* to journey). Deterministic chaos *bifurcates* (splits or divides) into *oscillating* dynamics (going back and forth between possibilities) that continue to bifurcate and oscillate, leading to cascades of bifurcations and oscillations. When graphed by a computer, these dynamics produce phantasmagoric images called *fractal attractors* (see below).

The bifurcations and oscillations of deterministic chaos iterate and mutually inhibit one another. However confusing this all may sound, don't despair. It simply means that the oscillations and bifurcations feed back into the dynamic (iterate), all the while moderating (mutually inhibiting) one another. We can describe this through an equation such as:

$$y \rightarrow ry(1-y)$$

where $0 < y < 1$

When the value of y is between one and zero, as y increases, $(1 - y)$ decreases. Correspondingly, as y decreases, $(1 - y)$ increases.[45] Thus, the equation expresses how aspects of the system influence one another. The

[45] For example, if $y = \frac{3}{4}$, then $(1 - y) = \frac{1}{4}$; if $y = \frac{1}{4}$, then $(1 - y) = \frac{3}{4}$

arrow denotes that the result of each computation is fed back into the equation as the new value of *y*, whereupon the equation is recomputed, the result fed back into the variables, the equation recomputed, and so on ad infinitum. We'll get to the letter *r* in a moment.

The equation portrays a bifurcated system that oscillates with itself. When that oscillating system in turn bifurcates, we have two systems that oscillate not only with each other but also within each other. Oscillations are now oscillating with oscillations. Furthermore, new bifurcations and oscillations occur with every iteration of the system. Please note that the bifurcations and oscillations are a result of the inner workings of the system and are not introduced from outside. That distinction is important for our accounts in Part Two of how to work therapeutically with deterministic chaos that is no longer functional.

Example: Planning a Vacation

Consider what it's like to plan a vacation with a friend. You suggest New York. Your friend suggests Chicago. A bifurcation (New York and Chicago) has appeared in your planning. Now suppose that neither of you wishes to accede to the other. Each of you tries to persuade the other. Now the bifurcated system oscillates (between New York and Chicago).

One of you now attempts to compromise by suggesting New Orleans. The other, not taking to that idea, suggests Miami. The system bifurcates again, now between two sets of oscillations: between New York and Chicago and between New Orleans and Miami. Inevitably, comparisons between New York and New Orleans, New York and Miami, Chicago and New Orleans, Chicago and Miami begin to form. Bifurcations multiply and oscillate.

But something else may happen. You may try to influence each other by citing examples of the advantages to be found in each city. You may cite the Art Institute of Chicago. Your friend may counter with the Metropolitan Museum in New York. Yet another bifurcation that oscillates!

In response to your friend who cites the Met in New York you may say that you prefer the Guggenheim. Now, not only are bifurcations between cities oscillating but bifurcations within the cities as well. Everything becomes inordinately complex. The discussion turns chaotic.

Which nobody enjoys.

The *r* Factor

The rapidity of a system's bifurcations is denoted by the letter *r*. It determines how much energy is in the system, so to speak. We shall have much to say about this in our applications of chaos theory to analytic practice.

As bifurcations oscillate with one another and bifurcate further, patterns that are astonishing in their complexity begin to emerge from the chaos. Intriguingly, many of these patterns, called *fractal attractors,* resemble patterns in nature. Resembling themselves across scale (scale-invariant self-similarity), they never resolve into boundaries between oscillations. They defy definition. They boggle the mind.

For example, here is a computer-generated graph of a fractal attractor:

To see what deterministic chaos looks like in the real world, compare it with the following photograph of galaxy M101.

Fractal attractors portray how the iterations of a chaotic dynamic generate ongoing patterns. We can see these patterns expressed pictorially thanks to the development of computers, which only recently have been able to graph them. Such graphs manifest patterns that endure while the dynamic continues to iterate.

The pattern is fractal because it looks the same no matter what the magnification (self-similarity across scale) and its dimensions do not conform to integers (fractal dimension).[46] *Fractal* describes how chaotic attractors transform the typical characteristics of non-chaotic systems.

Another characteristic of chaotic systems is SDIC, which stands for *sensitive dependence on initial conditions*. Oscillating bifurcations that bifurcate into further oscillating bifurcations are exquisitely sensitive to where they start. Where they end up has much to do with where they begin. Thus, even the smallest difference between origins produces profound differences as the system iterates.

---

[46] Think of a line having a dimension of one (length); a plane having a dimension of two (length and width); and a solid having a dimension of three (length, width, depth). Fractal dimensions fit in between these integers. Thus, the fractal dimension of an irregular solid could be represented by a number like 2.7.

Perhaps the most familiar example of SDIC is the "butterfly effect." The metaphor is of a butterfly flapping its wings in China. The resulting disturbance of air, although infinitesimally small, disturbs the air around it. As thermal influences interact with those disturbances, the effect of the butterfly's wings is amplified and multiplied as the air mass travels eastward on the jetstream. Eventually, as the original disturbance of air reverberates through minutes, hours and days, constantly feeding back into the atmospheric dynamics it encounters, profound atmospheric disturbances develop. Eventually, there are thunderstorms in Omaha.

That butterfly in China had quite an effect on Omaha. Flapping its wings iterated in feedback loops that multiplied so rapidly that chaos resulted. That the dynamic is chaotic means that it is also self-similar across scale. Were we able to freeze-frame the atmosphere at any one of its iterations beyond the original flapping, we would find that each successive iteration preserves the character of the original. The dynamics are consistent across scale.

They are also so complex that they cannot be differentiated into discrete categories. In fact, they transcend categories. Their dimensions are way beyond the familiar categories of one, two, three, and four (line, plane, solid, time). They occupy every possibility in between. Rather than being measured by single integers, they have dimensions like 1.6, 2.4, 3.1, 4.3.

SDIC, self-similarity across scale, and fractal dimension are presented in greater detail in *Archetypes*. For now, we need only to know that these elements exist. They will become very important for our understanding of how chaos enters the consulting room.

Finally, fractal attractors lend excellent visual expression to the complex inner workings of the brain.[47] Could they also illuminate much of the science of higher mental processes? From the perspective of Jung's metapsychology, they can.

---

[47] For an account of Walter J. Freeman's discovery that fractal patterns encode the olfactory bulb's responses to odors, see my *Archetypes,* pp. 65ff.

Analytical Psychology

As explained in detail in *Archetypes,* Analytical Psychology parallels chaos theory in its descriptions of the structure and dynamics of the psyche. For example, conscious and unconscious endlessly feed back into one another through perceptions and behaviors. In fact, perception and behavior themselves form a dynamic that iterates and mutually inhibits as it bifurcates and oscillates, for every behavior influences the outcome of a situation, which in turn influences perceptions of future situations, which then influence successive behaviors, and so on throughout life.

Jung told an interesting story of how events can condition perception and behavior. He was walking at dusk with a man who, upon hearing the church bells, complained that he could not stand the noise of their ringing. Jung felt this to be rather curious and, upon further examination, discovered that the man had had an unpleasant encounter with the pastor of the parish. Although this had happened many years ago, it conditioned the man's experience of the bells.

From the perspective of chaos theory, the man's emotional reaction was a function of SDIC. His unpleasant experience with one particular man developed into annoyance at church bells in general. As his experience resonated through the years, it remained self-similar across scale. No matter where or when he heard them, his anger at the bells was identical to his anger with the pastor. Finally, his annoyance manifests fractal dimension, for there is no obvious link between bells and people. It transcends categories.

Engaging the Feedback Loops

Perceptions and behaviors routinely transcend logic and rationality. They can be as complex as chaotic dynamics and fractal attractors. Attempts to decipher the chaos of psychological processes through logic and rationality succeed only to a degree—and a small one, at that. How, then, can we work with unconscious processes beyond that which is possible through rationality?

For Jung, the answer lay in *active imagination,* which consists of dialogues between conscious and unconscious. The unconscious consists of synaptic configurations formed around life experience. The beauty of active imagination is that it allows us to engage consciously any one of these configurations. That adds energy to the configurations, which then amplifies their activity in ways that can make them more visible to consciousness. Holding unconscious configurations in consciousness intensifies the interactions between the two. The resulting bifurcations multiply into further oscillations.

As conscious and unconscious continue to interact, their bifurcations and oscillations increase and diversify. Thus, active imagination raises the *r* factor between conscious and unconscious. If the tension between conscious and unconscious is incubated (that is, held in tension without attempting to seek resolution), a *uniting symbol* eventually appears.

Uniting symbols consist of a third possibility that resolves the conflict between conscious and unconscious by constructing a bridge between them. Remember, this is an oscillating system. Thus, when consciousness recognizes the uniting symbol, the unconscious poses an alternative, which establishes a new conflict. Once consciousness recognizes that situation and re-engages the unconscious, the potential for a new uniting symbol emerges.

In other words, oscillations between conscious and unconscious generate bifurcations that oscillate with each other. These bifurcations and oscillations form the core of—and are essential to—individuation.[48] Here it is in Jung's words:

> This process [constellation of opposites and subsequent resolution via "the incommensurable third"] can … take place spontaneously or be artificially induced. In the latter case you choose a dream, or some other fantasy-image, and concentrate on it by simply catching hold of it and looking at it. You can also use a bad mood as a starting-point, and then try to find out what sort of fantasy-image it will produce, or what image expresses this mood. You then fix this image in the mind by concentrating your attention. Usually it will alter, as the mere fact of contemplating it animates it. The alterations must be carefully noted down all the time, for they reflect the psychic processes in the unconscious background, which appear in the form of images consisting of conscious memory material. In this way conscious and unconscious are united, just as a waterfall connects above and below. A chain of fantasy ideas

[48] For a more thorough explanation of this process, please see *Archetypes,* pp. 26 ff.

,develops and gradually takes on a dramatic character: the passive process becomes an action.[49]

Jung believed that active imagination is essential to both individuation and the analytic process. Interpretation, which defines and clarifies unconscious contents, is an excellent first step that differentiates images from the unconscious. Moreover, it lends the analytic process an atmosphere of recognizability that both analyst and patient often find reassuring. Nevertheless, conscious clarification is only one aspect of the analytic process. If not supplemented by active imagination, it can actually hamper individuation.

How so? Whenever we categorize the amorphous contents of the unconscious through clarification and definition, whatever doesn't fit gets left out. Rather than increasing our contact with the unconscious, we decrease it. Better to keep the dynamic between conscious and unconscious unfolding through bifurcations and oscillations.

Consider how Jung describes the process in the above passage: "concentrate on it by simply catching hold of it [a dream, or some other fantasy-image] and looking at it," and "fix this image in the mind by concentrating your attention." Remember r in the above equation for deterministic chaos? By holding the image in mind we intensify (r in the above equation) the tensions between conscious and unconscious. In the language of chaos theory, we increase the intensity and frequency of the oscillations between the two. Or, in Jung's terms: "contemplating … animates."

Raising r, forces the dynamic to bifurcate. The resulting chaos generates "a chain of fantasy ideas [that] gradually takes on a dramatic character." From the perspective of chaotic dynamics, such fantasy chains are fractal. Eventually, they form fractal attractors, that is, configurations into which the dynamics periodically settle—uniting symbols, in Analytical Psychology terminology.

Thus in this way the analytic process raises the intensity (r) of the oscillations between conscious and unconscious—$y(1 - y)$—as they iterate and intensify by feeding them back into each other. The result is

---

[49] "The Conjunction," *Mysterium Conjunctionis,* CW 14, par. 706.

that the original fantasy image "usually…will alter, as the mere fact of contemplating it animates it."

"I THINK YOU SHOULD BE MORE EXPLICIT HERE IN STEP TWO."

That the unconscious generates uniting symbols and fantasy images can seem miraculous because we have very limited access to the fundamental mechanics of the analytic process. Conscious and unconscious are by definition largely incompatible. Like deterministic chaos, the complexity of the inner workings of the psyche frustrate definition and explanation.

Unconsciously produced images transcend logic and rationality. Symbols cannot be reduced to particular meanings.[50] Consequently, Jung took a constructive approach to interpretation. He maintained that understanding the meaning of a symbol could be achieved only by looking at it wherever it has appeared, both individually and collectively. If in

---

[50] Actually, Jung did feel that one type of symbol did have a consistent meaning. Water usually refers to the unconscious, particularly the collective unconscious.

dreams, we begin with the dreamer's associations; if in history, with folktales and myths.

Fractal attractors and symbols resemble each other. Neither can be reduced to specific analyses of their dynamics. Their images are exactly what they are, and nothing else. They do not refer to something else, nor can their complexity be deconstructed into simpler fragments that define the whole. They simply are what they are.

Furthermore, the unconscious dynamics that symbols reflect profoundly affect consciousness. Simply wondering about the meaning of symbols resonates with the unconscious. As with iterative equations, the result of every computation influences the outcome of the next iteration. Every time we consciously engage a symbol, the unconscious responds. Likewise, whenever the unconscious presents us with a symbol, we consciously respond. Like y and $1 - y$ in the mutual inhibition equation, conscious and unconscious modify each other.

Where To From Here?

In the next chapter we'll explore how chaotic dynamics inhabit the psyche's attempts to integrate expectation with actuality. In defending ourselves from the assaults of the environment that threaten our stability, we unwittingly preserve the expectations of instability through the very means by which we attempt to finesse it. Consistent with what we have learned about the chaotic nature of individuation, we'll see how it stalls when we try to dampen the oscillations between conscious and unconscious. The result is neurosis, which frustrates development until that which blocks us is recognized and transformed.

Once we accept that complexity is the engine of development (however it may frustrate conscious control), individuation can proceed. We gain the means to tolerate confusion and uncertainty as we participate in a process whose ends are unpredictable. We may even be able to perceive the logic of the psyche's development.

In the next chapter we'll focus on paradox. We'll call the ways in which the unconscious deals with threatening childhood environments *defenses*. We'll examine how defenses bifurcate and oscillate by simultaneously protecting us from and exposing us to threats to our psychological stability. We'll then look at how self-defeating behaviors and perceptions evolve from defenses as previous solutions to essentially unsolvable problems become the new problems.

Chapter Five completes our theoretical elaboration by looking at trauma. Trauma is a good example of how good solutions can go bad, for trauma symptoms preserve the original trauma. That which aims to relieve actually exacerbates.

To recognize, and then to tolerate, the chaos that generates and enlivens individuation is to progress beyond the neuroses that block our development. We must learn to appreciate how paradox alerts us to the chaos that calls to us from just beyond our conscious abilities. Through it we can discover and engage the fullness of our being. Only then can we become who we are meant to be and most satisfyingly are.

4
The Paradox of Defenses

Analytical Psychology and chaos theory share a common theoretical foundation, namely that both describe dynamics that are characterized by tension. Jung believed that conscious and unconscious constantly interact in ways that sometimes provoke, sometimes amplify, yet always compensate each other. Sometimes consciousness saves the day; at other times that role falls to the unconscious.

For example, the unconscious reacts more quickly than consciousness to external events. At times, this can be very adaptive. If I step off the curb and hear a car horn, my unconscious is likely to rescue me much more rapidly than consciousness, which requires so much time to analyze the situation and to decide what to do that I may be dead before I react.

Consciousness also has its advantages. If my unconscious tempts me to blast my horn at the driver in front of me, who seems never to have grasped the significance of turn signals, my consciousness may inhibit that action on the basis that it will only aggravate the situation. Or, if my unconscious redirects my affections from my partner to a more compelling other, my consciousness may remember that we've taken that road before with less than satisfactory results.[51]

The unconscious is very good at bringing up what consciousness forgets or ignores. For example, if I am attending a job interview, I may effect an attitude of confidence and optimism in order to impress my interviewer(s). While that may help me to win the job, or pass an oral exam, or clear through customs, my effected confidence is only part of the psychological story.

That night, I may dream exactly the opposite: That I do not get the job, or that I fail the exam, or find myself repatriated. The unconscious is

---

[51] Such redirections of psychic energy can be important generators of psychological development if we consider that our attractions actually refer to underdeveloped aspects of ourselves. The unconscious routinely directs our attention to anything—or anyone—that embodies such aspects. We'll have more to say about this in the final chapter.

not sabotaging us. It is simply reminding us of what was left out during waking hours. That we may not wish to include our most negative and pessimistic feelings in the face of a challenge does not make these feelings go away. They are simply suppressed. Dreams return them to center stage.

Unconscious intervention occurs especially when consciousness is inadequate to a situation. If my job is literally killing me, my unconscious may react in ways that make it impossible for me to go to work. I may become depressed, or I may develop a physical symptom like ulcers, hypertension, or carpal tunnel syndrome. Such unconscious reactions serve to avoid stress in order to improve our health.

Conscious and unconscious complement each other. The one cannot do without the other, particularly because both adapt to life's challenges in ways that can become maladaptive. Our concern is with those that develop before consciousness is functional, at a time when it is up to the unconscious to adopt strategies (perceptions and behaviors) that facilitate survival. Later on, those very coping strategies may themselves become problematic. Then, the unconscious needs the reflective and analytic capacities of consciousness to find its way out of the dilemma.

That the relationship between conscious and unconscious bifurcates and oscillates enhances survival. Such tensions of opposites characterize the kind of chaotic dynamics that can increase options and choices. For example, in *Archetypes* I cited the work of Ari Goldberger, who describes how chaotic dynamics in the sinusoidal wave of the heartbeat enhance survival when the heart becomes arrhythmic. The more frequencies available for the heart to restore its equilibrium, the better its chances of surviving an insult.[52]

The Paradox of Defenses [53]

Paradoxes embrace contradictory elements. Thus do bifurcations and oscillations between conscious and unconscious manifest themselves as

---

[52] See A. Goldberger, D. Rigney, and B. West, "Chaos and Fractals in Human Physiology," *Scientific American,* vol. 262, no 2 (February 1990).

[53] For a more complete elaboration of the paradox of defenses, see my paper, "The Analysis of Defenses," in *The Journal of Analytical Psychology,* vol. 36, no. 2 (April 1991).

paradoxes. While that may leave us scratching our heads, simply recognizing that a paradox exists is the first step to facilitating its role in individuation.

The chaotic relationship between conscious and unconscious can be quite disturbing. Even though our egos evolve out of unconsciousness, we are more comfortable with the ego's activities. Reflection, contemplation, analysis, and choice seem much more stable than the seemingly impulsive activity of the unconscious. Nevertheless, there is much to be learned from the chaos that unconscious processes generate.

Consider paradox, which requires that we hold two seemingly contradictory things in mind. That can feel far less efficacious than the single-minded determination of consciousness. But Jung thought paradox had a very useful role in individuation:

> Paradox … does more justice to the unknowable than clarity can do, for uniformity of meaning robs the mystery of its darkness and sets it up as something that is known. That is a usurpation, and it leads the human intellect into hybris by pretending that it, the intellect, has got hold of the transcendent mystery by a cognitive act and "grasped" it. The paradox therefore reflects a higher level of intellect and, by not forcibly representing the unknowable as known, gives a more faithful picture of the real state of affairs.[54]

Jung dedicated his career to exploring the value of the unconscious. He held that consciousness had to learn to be comfortable engaging that which does not fit its categories and specificities—the "darkness" of "transcendent mystery." Paradox, which respects that mystery, "reflects a higher level of intellect" that "gives a more faithful picture of the real state of affairs" than does either "clarity" or the "human intellect."

That may explain much of the appeal of cognition-oriented therapies, for the presumed clarity of intellect usually feels much more reassuring than the darkness of transcendent mystery. Only in the latter, however, do we gain the raw material for intellect's analysis. But intellect cannot always endure the paradox of mystery.

---

[54] "Transformation Symbolism in the Mass," *Psychology and Religion: West and East,* CW 11, par. 417.

Perhaps nowhere does the Janus-faced character of paradox predominate more than in defenses. Many might wonder why we are looking at defenses at all, for Jung rarely mentioned them. When he did, he seemed to conceive of them primarily as a means by which the ego protects itself from the intrusive power of the unconscious, which is strongest during childhood. For example, he says:

> Consciousness did not exist from the beginning; in every child it has to be built up anew in the first years of life. Consciousness is very weak in this formative period, and the same is true of the psychic history of mankind—the unconscious early seizes power. These struggles have left their mark. To put it in scientific terms: instinctive defense-mechanisms have been built up which automatically intervene when the danger is greatest, and their coming into action during an emergency is represented in fantasy by helpful images which are ineradicably imprinted on the human psyche.[55]

At first glance, defenses seem simply to protect the integrity of the ego, intervening "when the danger is greatest." Yet if they generate "helpful images" as well, then defenses must play a dual role: to protect the ego and to contribute to its development. While such an idea is not developed by Jung, it conforms to a fundamental premise of his metapsychology, that of the potentially proactive and positive role of the unconscious.

Jung agreed with Freud that:

> the patient remembers nothing of what is forgotten and repressed, but … expresses it in action. He reproduces it not in his memory but in his behavior; he repeats it, without of course knowing that he is repeating it. For instance, the patient does not say that he remembers how defiant and critical he used to be in regard to the authority of his parents, but he behaves that way towards the physician.[56]

What occurs in the transference reflects what occurs in life. Both Freud and Jung believed that children's consciousness is not adequately developed to recognize, ponder, consider, or analyze early life experience. Rather, those experiences and the response of the organism become

---

[55] "Psychotherapists or the Clergy," Ibid., par. 533.

[56] Sigmund Freud, *Therapy and Technique,* p. 160.

encoded in the unconscious, where they appear later in life as perceptions and behaviors that may or may not be relevant to adult life experiences. Jung especially saw defenses as highly complex psychic structures that not only protect us from repressed memories but—and here the language of chaos theory is particularly helpful—bifurcate and oscillate, as well. By highlighting potential threats in the present, they bring to light troublesome dynamics in the past.

The Chaos of Defenses

That itself is a two-edged sword. By bifurcating and oscillating between past/present and threat/protection, defenses conjure up past disasters and offer opportunities to address them through engaging their remnants in the present. In the language of chaos theory, defenses reveal fractal dimensions of unconscious processes formed earlier in life.

Defense-generated symptoms usually enter the analysis via the transference and countertransference. The goal of analysis is to welcome them into the analytic relationship, where they can be brought to consciousness and transformed into more functional behaviors. Generally, analysts should forego interpretation in favor of amplification so that the feedback loops that result can intensify the oscillatory dynamics of the defense, thereby raising its *r* value.[57]

The fractal dimensions of defenses offer insight into the conflicts that generated them in the past and to which they attach themselves in the present. Thus, defenses not only defend against past conflicts, they give access to them as well. They dilute the threat and empower healing. The challenge for analyst and patient is to engage psychic processes that are at first only vaguely understood. The very chaos sown by these processes is the healing element within the situation, the key to which lies in feeding back into the unconscious a response from consciousness. Thus, patients must enter into the chaos of their defenses in order to create feedback loops. This, rather than the validity of the analyst's interpretations, allows them to discover what the defenses are all about.

Defenses are not simply the vehicles of suppression that many generic

---

[57] *Archetypes,* p. 18.

definitions suggest: "In general, psychological defense mechanisms modify, distort, or render unconscious the feelings, thoughts, and memories that would otherwise provoke anxiety."[58] Such definitions assign to defenses the role of exerting a dampening effect on the "feelings, thoughts, and memories" that become activated by "internal psychological threats."[59]

"Internal psychological threats" do not simply arise *de novo*, however. They originate in our responses to threatening situations in the environment. We don't manufacture feelings of being threatened. We recognize them on the basis of past experience. Trying to dissolve a defense through minimizing the threat usually meets with little success, for

> even when there is no objective danger, the appraisal of a situation as threatening will evoke an emotional reaction proportional in intensity to the magnitude of the perceived danger or threat. Thus, whenever an event or situation is seen as threatening, irrespective of whether the danger is real or imagined, the sense of threat will lead to an unpleasant emotional reaction—the arousal of an anxiety state.[60]

As we shall see in our description of trauma, this is the origin of flashbacks.

That is how defenses selectively condition perception to be sensitive to certain aspects of experience. While the threatening element may comprise only a very small dimension of a particular situation, the focus of the defense will preoccupy our attention to such a degree that the small threat may characterize the entire experience.

Consequently, defenses that evolve to protect us from a threat inevitably sensitize us to the potential presence of that threat. They create a proximity with that which generates anxiety. Yet, by consistently bringing us closer to that which is perceived as a threat, defenses create opportunities both to reevaluate the nature of the threat and to modify our responses to it.

---

[58] Samuel M. Turner and Michel Herren, eds., *Adult Psychopathology and Diagnosis,* p. 280.

[59] Ibid.

[60] Ibid., p. 275.

We are particularly interested in defenses that originate in early childhood, before consciousness is fully awake. These can be especially difficult to integrate with consciousness, having never been conscious to begin with. They also tend to be problematic, for however adaptive they were originally they are less so in adulthood. They may squander our energy on threats that are actually quite innocuous. Or they may lack the sophistication to cope adequately with threats that are formidable. Either way, individuation is impaired.

Jung defined this state of affairs as *neurosis*, wherein psychic energy is blocked until it regresses (oscillates) to that time in our development when the unconscious implemented adaptations that are that are no longer functional When we find ourselves in this psychological past, we can consider the choices that, while not adaptive then, may be exactly what is needed now.

By allowing psychic energy to regress into the unconscious, where archetypal images can be activated, the ego sacrifices itself to a process that recapitulates the past. In effect, the ego exposes itself to the shadow. Analysis engages this process as consciously as possible, so that understanding can beget volition.

As adults we cringe at the thought of acting like children. Nevertheless, our childhood defenses take us there anyway. Defenses serve "the polaristic structure of the psyche," through which tensions of opposites generate and direct the flow of psychic energy toward psychological growth, or individuation.[61] What may seem like an unfortunate eruption of childish emotion is actually an opportunity to integrate what was left behind.

"Polaristic" means that everything in the psyche has a shadow side. This duality generates tensions of opposites, without which growth is simply impossible. Thus Jung says, "The shadow and the opposing will are the necessary conditions for all actualization."[62] As the generator of tensions of opposites which balance the psychic organism, the shadow

---

[61] "On the Psychology of the Trickster-Figure," *The Archetypes and the Collective Unconscious,* CW 9i, par. 484.

[62] "A Psychological Approach to the Trinity," *Psychology and Religion,* CW 11, par. 211.

archetype is a creative urge within the Self. Like Lucifer (the "Light-Bringer"), it prevents the universe that is the psyche from ever being "just a piece of clockwork which the Creator has to wind up to make it function."[63]

Recognize the chaos in all this? It's a cornerstone of Jung's metapsychology: Creativity arises out of the chaos that ensues when opposites pull at one another. The bifurcations and oscillations native to psychic functioning are essential to individuation. Without them, we'd be stuck in a linear existence, the direction of which would be immutable. Life would be one endless rut. Change would be impossible.

Fortunately, the psyche is an endless cascade of bifurcations and oscillations between conscious and unconscious. The ensuing fragmentation of the ego is an essential part of individuation.[64] Old identities must die—must be sloughed off, so to speak. For all life forms this is a painful process that involves a great deal of effort, for the struggle to wrest the new from the old always meets with resistance, if only from inertia. Whatever suffering ensues is necessary and positive, for transformation requires a leaving off of the old to prepare the way for the emergence of the new. No butterflies without the sacrifice of the caterpillar.

Playing a central role in this process is the archetypal shadow.[65] A dynamic that functions from birth, it is the engine of chaos, balancing and unbalancing the psyche so that change and growth can occur. Adaptability is enhanced when we become aware of this process. Becoming conscious of the tensions and conflicts that exist in the unconscious (primarily through dream analysis, but also through active imagination and fantasy) allows us to participate actively and creatively in the creative growth of the psyche.

Bifurcation and oscillation are endless. Tensions and conflicts that exist in consciousness have their counterpart in the unconscious. Thus, conflict in consciousness begets resolution in the unconscious. Likewise,

---

[63] Ibid., par. 290.

[64] Jung believed that fragmentation of the ego was reflected even in alchemy, wherein elements must suffer fragmentation before they can be transformed into new wholes.

[65] See chapter four in *Archetypes*.

harmony and resolution in consciousness have their opposite (tension and conflict) in the unconscious. To become conscious of tension and conflict allows harmony and resolution to appear in the unconscious.

Once we recognize this process we can draw upon the helpful input of unconscious processes during times of duress and understand why times of quiescence are short lived. Most importantly, we can affirm the creative dynamics of the individuation process. We can grow beyond what we are. Liberated from the past, we can embrace the future.

The Creative Role of Defenses in Individuation

The key to working with defenses lies in recognizing how they amplify the psyche's inherent oscillatory dynamics. While Jung's use of metaphors of tensions of opposites has sometimes been criticized as being essentially dualistic, and therefore too restricted, ample evidence exists that the psyche is largely a function of oscillators. In fact, "neurons with oscillatory properties are a common feature of the nervous system."[66] Furthermore, as we have observed in deterministic chaos, tensions of opposites are not static entities. Rather, they are dynamic, altering whatever comes into their proximity and themselves being altered by the ensuing interactions. Tensions of opposites bifurcate into multiple oscillating systems. Thus does duality become plurality.

Jung described archetypes as *bipolar*. Andrew Samuels calls it "the bifurcated nature of the archetype."[67] It is precisely this bifurcating dynamic of archetypes that is recapitulated in defensive dynamics. Thus, defenses and archetypes are fractal, manifesting themselves in similar patterns over differing levels of functioning. Like archetypes, defenses bifurcate into tensions of opposites—those, for example, between present and past, consciousness and the unconscious, patient and analyst—that bring shadow contents into consciousness. Like trickster figures, archetypes and defenses challenge the ego's perspectives with material from the unconscious. When consciousness interacts with that material, the transcendent function brings the two into paradoxical congruence

---

[66] Thomas B. Kepler, Eve Marder, L.F. Abbott, in *Science,* vol. 248 (April 1990), p. 83.

[67] Samuels, *Jung and the Post-Jungians,* p. 28.

through uniting symbols. Hopefully, consciousness will expand. Too often, however, we avoid the whole thing by resorting to suppression.

Let us then propose a working definition of defenses, namely, that they are the mechanisms by which unresolved conflicts are brought to consciousness in a manner that is diluted enough to maintain the integrity of the ego while charged sufficiently to empower its response. Thus, as with Jung's definition of the psyche as bipolar, defenses both dilute the threat and empower transformation. They are, in the words of the previous section, paradoxical.

Much depends on which capacity of the defense we choose to reinforce: dilution or transformation. It helps to be able to link the activity of a defense with former conflicts. It is here that the metaphors of chaos theory prove especially helpful, for defenses function exactly like fractal attractors. They establish patterns that, while clearly repeating themselves, rarely ever retrace the same paths. Furthermore, they utilize oscillatory dynamics to create basins of attraction, the boundaries of which can never be reduced to a single element, but constantly open up, under further scrutiny, into ever greater complexities of interaction. Finally, their dynamics can be described by mutual inhibition equations. By raising the r value we can intensify the bifurcations and oscillations by adding to them our own, thereby generating new potentials for adaptation.

To understand these dynamics is to gain valuable insight into utilizing defenses in the service of individuation. In the next chapter we'll look at how the unconscious's response to challenges to survival in early childhood can lead to defenses that become less than functional later in life. Remember, until consciousness is sufficiently up and running, the unconscious runs the show. If we can separate ourselves sufficiently from the presumed security of consciousness, we can perceive the brilliance of the solutions the unconscious implemented early in childhood. If those solutions later become problems, however, dialogues between conscious and unconscious must come into play.

Because challenges to survival are always traumatic, we'll begin the next chapter with a brief review of trauma theory. We'll look at three primary symptoms that can lock in early childhood responses to trauma, which often compromise our adaptability. Then, on to Part Two, where we'll generate some new metaphors that can assist us in unlocking the childhood defenses and examine some examples of how to redeem them.

5
The Strange Attractors of Childhood Trauma

Imagine what it must be like to be a very young child living in a war zone. Danger surrounds us. At any moment our lives could end. The adults are powerless to do anything about it. There is, in fact, nothing that can come to our aid. We live a life of continual shock, as bombs fall indiscriminately, people are killed for no apparent reason, and every necessity of life is at best in short supply, at worst, completely unavailable.

Or maybe we have been born into a drug-addicted family. Father left long ago. We'll never know him. Mother is incapacitated, passed out in a drug stupor day after day. We'll never really know her either. No one looks after us. No one holds us, no one feeds us, no one comforts us, no one seems even to care. We are alone in a dark world of ceaseless neglect.

Or maybe we are born into a reasonably functional family. Dad is still around, but he is away most of the time trying to find work. Mom has to work, too, so there are extended periods when neither is around. Fortunately, there is an uncle who can look after us. Or maybe it's a grandmother. Unfortunately, due to whatever distorted childhoods they suffered, they are desperately trying to connect intimately to the child inside that they had to leave behind in order to survive. They mistake us for that child and repeatedly connect intimately with us. Serially abused sexually, we live in a sickening world of utter confusion.

Or maybe we fall in love with a girl in our seventh-grade class. This has never happened before. In fact, not too long ago we found girls a bit repugnant. We're terrified and confused. What should we do? Her image torments us. Life feels dead without her. Finally, one day, we can stand it no longer and we say "hello" to her. But our timing is bad. She is just about to leave with the handsomest boy in class, who is just about to take her by the hand. She looks at us as if she's never seen us before, as if she knows she has to be polite, and says in passing: "Hi." The boy smiles a sneer at us as they pass. As we turn away, a shy girl who has a crush on us says: "Hello!" We walk past without answering.

We could go on. But the point is clear. We can't deal with such situations. Unfortunately, we're also on our own and our consciousness is not sufficiently developed to sort things out, to reflect on the situation, and to formulate some sort of plan for survival. For all appearances, the situation is hopeless. We have nothing with which to cope. We're out of luck.

At least that's how it feels. But we are not alone. We do not lack resources. Deep inside our developing brains is a limbic system up and running, ready to get us through whatever we face as best it can. The billions of synapses we were given at birth are already factoring in the challenges to our survival. Like the parasites we read about in Chapter One, synaptic biochemistry does whatever it needs to get us through.

If it's successful, we'll live with those adaptations until someday we are conscious enough to become aware of them. But they're unlikely to appear as felicitous mantras in some sort of psychological Disney World of life.

No. They're more likely to appear as conflicts with our adaptations to adult life. We'll see them as maladaptive. They'll cause us no end of trouble. We'll end up hating them. We'll seek psychotherapy, medication, institutionalization, maybe even suicide, all in pursuit of getting rid of them.

As with the paradoxical nature of defenses, it's exactly the wrong thing to do. Those synaptic configurations continue to fire off when we are adults. They are effectively the remnants of who we were when we were barely conscious children. That little child in us who managed to find a way to get us through the horror of violence, neglect and abuse is now receiving the same from us. We do all we can to ignore or eliminate them. Rather than the defense, we ourselves are now the threat to our survival. The sooner we realize that, the sooner we can be on the road to healing.

Synaptic Plasticity

Traumatic experiences are sometimes called "flashbulb" experiences due to the suddenness with which they intrude on our lives, demolishing in an instant the illusions of control that make it possible to face the day. Auto accidents, "Dear John" letters, natural disasters, layoffs, and the like are single episodes that leave long trails of distress.

The first three above examples, however, are hardly that. Those are long-term chronic experiences that leave us more than traumatized. They leave us permanently damaged, eternally coping, always one step from misery. The reasons for that are many, but we shall focus on two. Firstly, our bodies are programmed from birth to cope with threats to our existence, be those acute or chronic.

Secondly, those coping mechanisms leave traces in the form of synaptic configurations that do not simply go out of existence. They become active whenever life presents us with anything that even remotely resembles the original trauma. We'll focus more on this second phenomenon in the rest of the book. For now, we'll briefly outline an important way in which physiology copes with trauma.

Obviously, this is not a comprehensive review of the biochemistry of the brain. It simply sets the stage for Jungian-style clinical interventions. We shall be concerned less with comprehensively explaining the complexity of the physiological responses to trauma than with simply noting that they exist. Our concern is with the residues they leave in the psyche in childhood that become the focus of therapy in adulthood.

Much is being done to track the physiological sequelae of ongoing trauma in childhood. Of special concern is the relationship between trauma and hormonal activity. For example, cortisol is a glucocorticoid that becomes elevated when we are in distress. It has been discovered that "children whose early morning cortisol levels vary a good deal from day to day, often spiking to higher than expected concentrations, had mothers who were clinically depressed during the child's infancy."[68]

Abandonment is traumatic. When we are very young, we lack the cognitive capacities to understand why we are being abandoned and what we might be able to do about it. We simply suffer in a wilderness devoid of landmarks and navigational aids. But our bodies are not helpless.

---

[68] J. Herbert, S.L. Halligan, I. Goodyer and L. Murray, "Exposure to Postnatal Depression Predicts Elevated Cortisol in Adolescent Offspring" (2004), in *Biological Psychiatry,* 55 (4), pp. 376ff; cited in M. Gunnar, "Stress Effects on the Developing Brain," in D. Romer and E. Walker, *Adolescent Psychopathology and the Developing Brain: Integrating Brain and Prevention Science* (2007), p. 138.

They immediately begin to mobilize the means by which to cope with the situation. Unfortunately, the chemical means by which the brain attempts to discern the context of the trauma and to cope with it can work at cross-purposes.

Research has shown that "the exact conditions that lead to a weakened ability to form explicit memories, and to regulate fear by thinking and reasoning, can also amplify fear reactions and enhance our ability to implicitly store information about stressful or traumatic situations."[69]

There's good and bad news here. The good news is that, even when the ability to form explicit memory is impaired, we can store useful information about harmful situations. The bad news is that, if we don't know what it is we are learning about, those stimuli might on later occasions trigger fear responses that will be difficult to understand and control, and can lead to pathological rather than adaptive consequences.[70]

These responses that will be difficult to understand and control, and can lead to pathological rather than adaptive consequences, are our main concern. It is exactly these kinds of dynamics that become programmed into our psyches that cause problems later in life. Yet, if we can become aware of them when they are active and can adopt a relational attitude toward them, healing can occur.

Our first step is to convert feelings into images.

But let's return to biochemistry for a moment. Right after birth,

> There is a period when the HPA axis[71] is relatively quiescent but this period is under strong social regulation. When the child is receiving sensitive, responsive, supportive care, the HPA axis remains quiescent, even when the child is behaviorally upset and distressed. When that care is removed and not immediately replaced by similarly high-quality care from a substitute caregiver, the HPA axis becomes highly responsive to stressors.[72]

---

[69] Joseph LeDoux, *Synaptic Self: How Our Brains Become Who We Are*, p. 110

[70] Ibid., pp. 110-111.

[71] The hypothalamus-pituitary-adrenal axis plays a role in the production of glucocorticoids, which influence how the body responds to stress.

[72] Gunnar, "Stress Effects on the Developing Brain," in D. Romer and E. Walker, *Adolescent Psychopathology and the Developing Brain,* p. 136.

In other words, children are born with a biochemical apparatus that is prepared to become activated if stress levels become too high, but which will remain quiescent if sufficient support is available. Consequently,

> In relationships in which the infant has experienced sensitive and responsive care, the presence of the responsive caregiver somehow blocks activation of the HPA axis. This is true even for babies who are expressing high distress—that is, even when the emotion component of the stress-emotion system appears to be activated by the stressor. In contrast, if the infant is with someone who is insensitive, unresponsive, or intrusive, that person's presence does not block activation of the HPA axis and significant correlations between behavioral distress and cortisol elevations are observed.[73]

Obviously the relational environment that children inhabit is critical for their development. That's not news. What is news is the nature and duration of the repercussions. Environments that either deprive or abuse children produce biochemical reactions that endure throughout life. That is our concern. That the biochemical reactions are what Jungians call the unconscious allows us to utilize all of the brilliantly creative ways of interacting with the unconscious that have characterized Jungian analysis as established by Jung himself. The following passage may put things into perspective:

> One way of conceptualizing the emotion component of the stress-emotion system is of a reverberating bidirectional system in which information about threat stimulates the central nucleus of the amigdala to orchestrate fear/ defensive responses, increasing the tone of stimulation from the LC[74] to the amigdala.[75] This increased LC input lowers the threshold for subsequent amigdala activation.[75]

Notice particularly that "the emotion component of the stress-emotion system is of a reverberating bidirectional system." Here we have an analog to the bifurcating/oscillating systems that characterize deterministic

---

[73] Ibid., pp. 134f.

[74] *Locus coeruleus*: a region in the brain stem where the norepinephrine system is located. Norepinephrine is a neurotransmitter that participates in vigilance and fear behavior.

[75] *amigdala:* the part of the brain where perceptions are translated into emotions.

[75] Gunnar, p. 130.

chaos. This passage refers to the manner in which traumatic experiences connect directly with the amigdala, which is now known to be the major player in the stress-emotion system.

That is critical to the course of therapy. Early childhood traumatizing environments create, whether through neglect or abuse, synaptic configurations in the brain that connect directly to the amigdala, and not to the higher cognitive functions. That leaves children less able *to regulate fear by thinking and reasoning* (in LeDoux's words, quoted above). While activation of the HPA axis does increase our ability to implicitly store information about stressful or traumatic situations, there's not much we can do with that information. When we are very young, our thinking and reasoning are not sufficiently developed to provide assistance. Traumatic stress retards them even further.

Like the tissue damage that shares its name, the word *trauma* comes from the Greek word for *wound.* Traumatic experiences rip open our souls and psyches. They are especially difficult to sort out cognitively. In childhood, they form synaptic configurations that connect more strongly with emotion (the amigdala) than with higher cognitive capacities, which leaves children in a state of heightened arousal and with diminished coping skills. That's consistent with our contemporary view of the sequelae of trauma.

Obviously, the sane thing to do is to ward off further damage. While there are many ways to do this, and they differ according to the individual, we can describe them under three general headings: Hypervigilance, hyper- (or hypo-) arousal, and flashbacks. Each of these sets of symptoms aim to protect us from threats. Over time, each itself becomes a threat—again, the paradox of defenses.

Hypervigilance

Hypervigilance is relatively easy to understand. Think of the folk saying: "Once bitten, twice shy." If some life event has torn us apart, we would be crazy not to monitor the environment to ensure that it never happens again. Unfortunately, like any defense, hypervigilance demands constant and endless wariness. It promises to protect, while at the same time hypersensitizing us to the possibility of threat.

Let's look at an example that could happen to anyone. Imagine that one day while waiting to make a left-hand turn a truck barrels into the rear end of our car. After spending several days in the hospital and several weeks recovering at home—negotiating with the insurance companies, losing time at work, preoccupied with relieving pain, relationships with family and friends disrupted—we finally regain normal functioning. But from then on, whenever we are waiting to make a left-hand turn, we automatically monitor our rear-view mirror to see if any trucks are approaching from behind.

Hypervigilance prepares us to take evasive action should there be a recurrence of the trauma. Like any defense, however, it also brings the threat of trauma into our awareness at the slightest appearance of anything even remotely associated with the original situation. As most traumatic experiences are exaggerations of normal, everyday life, that means we are constantly on the alert … hypervigilant, actually. Not only is it exhausting, but the physiological responses to threat accrue and complicate over time. Our cortisol levels alone skyrocket, leading to all kinds of secondary symptoms.

Letting go of hypervigilance is very difficult, for it is driven by the nonrational, emotional centers in the brain.

Hyperarousal

Complicating our attempts to cope with hypervigilance is the state of our arousal. To ceaselessly monitor the environment for any sign of imminent trauma is to be in a state of endless hyperarousal. It is impossible to stay calm. We are always on edge. Correspondingly, we may adopt a "head in the sand" approach, like an ostrich that cannot tolerate the threat of imminent trauma. We may simply "numb out" into a state of constant hypoarousal in the face of possible trauma: "Nobody moves, nobody gets hurt."

Either way, our normal way of being becomes supplanted by feelings, perceptions and behaviors oriented toward avoiding trauma. However adaptive that may be in the short run, over time it takes a heavy toll on our bodies and minds. We are not designed to be forever vigilant. We also need to relax, to have "down time," to allow our minds to wander freely into new ideas and experiences.

Flashbacks

Why are we so obsessed with avoiding further trauma? One would think that over time the threat of potential trauma would mitigate. Fortunately, that does sometimes happen. But traumatic experiences are rarely discreet. Rather, they are extensions and exaggerations of everyday life experiences that threaten to do us in. As a result, we often find ourselves experiencing the fullest extent of a past trauma in anything that resembles the original situation in which the trauma occurred. We call these flashbacks.

Flashbacks are particularly troublesome after rape. To be raped is to be forced to endure a sexual act that is aggressive and invasive from start to finish. Healthy sex is characterized by affection, caring, mutual enjoyment and participation—everything rape is not. But at the point of penetration, healthy sex also involves a moment of aggression and force.

While that can be a source of pleasure for both partners, after rape it can infect the entire sex act, shoving aside everything else. All our partner has to do is touch us affectionately and we relive the entire rape. We recoil in fear. We forget who and where we are. No longer is someone we love reaching out to us. We are having a flashback. We are being raped.

Flashbacks are a prime example of deterministic chaos. Hypervigilance leads us to doubt the security of the environment. We oscillate between fantasy and reality. Hyperarousal raises the *r* factor. Flashbacks preserve the original pattern that generated the chaos in every further iteration. Life becomes fractal. The oscillating system is closed, with nothing intervening to generate new patterns of perception and behavior.

At first this may seem dysfunctional. How can we escape the original trauma if self-similarity across scale and fractal dimension preserve the original trauma in every future experience? Is this chaos simply a trap?

Actually, no. That the original trauma is preserved in perceptions and behaviors gives us the opportunity to enter back into it and change the outcome. Even better, we do not have to relive the original trauma to transform its effect on us. Because everyday life events can recall the trauma, it can be relived in everyday events. For example, conflicts with relatives can contain remnants of an interpersonal trauma. Healing that trauma can then occur through discussing the conflicts we have with relatives. Thus, we can heal trauma without actually referring to it.

> Because ... well, hell, I thought I was beyond being bugged from my past, you know, I thought I had cemented myself forever from the years of my youth. … And see: since I had considered myself cut loose from the past I had seen no reason to guard that direction … all my beautiful fortifications … had been designed in accord with information indicating that the only dangers lay in *front*, *ahead* of me—and were fortifications, alas, quite powerless against even the meagerest offensive from the rear.[76]

When analyst and patient discuss problematic life situations, they enter together into the remnants of the original trauma. We shall have much to say about this in Part Two. For now, suffice it to say that when two chaotic systems interact, the resulting increase in complexity creates new attractors that can integrate the old. That is, analyst and patient create synergy simply by interacting with each other. Such synergy has the potential to transform them both.

Dissociation

Being in an ongoing traumatized state is unendurable. Particularly in childhood, trauma forces us to dissociate from ourselves and from others. Psychologically speaking, we separate the experiencing ego from the observing ego. Such phenomena as drifting to the ceiling and watching ourselves being attacked, worked on medically, or tortured are forms of dissociation. By not inhabiting ourselves (primarily our bodies) we distance ourselves from the pain.

Dissociation can be acute or chronic. If acute, our disconnection from ourselves lasts only as long as the traumatic event. If chronic, it lasts longer, even to our entire lifetime. That is another paradox of defenses. To dissociate is to abandon ourselves. We separate from that part of ourselves that feels so much pain that our entire organism is threatened.

This is especially poignant in cases of neglect. Children who are

---

[76] Ken Kesey, *Sometimes a Great Notion,* p. 64.

neglected by their parents experience excruciating pain. Imagine what it is like to be condemned to a soundless vacuum of unfulfilled longing for attention, affection, interaction, recognition, appreciation. What must it be like never to experience all the variations of love between parent and child?

Most of us know what it is like to throw a party, invite people for a certain time, and have no one show up. Even though we know that they are simply late and will be there, we cannot help but feel a bit of panic at the prospect of being abandoned. For far too many children, that is all they know.

Neglect is incredibly toxic. The unbearable feelings that it creates overwhelm children. Their own inborn capacity to love becomes such a liability that it threatens their survival. Their very health is in such danger that the risk of succumbing becomes very acute. Some do die. Others survive by putting as much distance between themselves and their pain as they can.

Such dissociation can last for years, decades even. It can become such a familiar part of our lives that we may not even be aware that it exists. No matter how successful our defense against pain may be, however, eventually it overtakes us and we are forced to recognize its existence. Then, we are faced with the daunting task of retrieving ourselves from the painful situation that gave rise to the defense in the first place.

Trauma and Survival

Here's what it looks like: If, when we are very young, we have no chance of receiving the love we need to flourish in life, we must either perish or find a way to endure the pain. One way to endure the pain is to dissociate from it and get on with life. We may seek what we need from others: Teachers, relatives, clergy, group leaders, imaginary relationships with stuffed animals or even figures in the media.

Or we may choose isolation and establish an identity as one who is alienated. We may then find others of like mind with whom we can bond (to the degree that it's possible). Or, we may become loners and never bond with anyone.

Either way, survival is purchased at the price of self-alienation.

Through it all, an abandoned child stares hollowly from the wings of the stage. That abandoned child is ourselves. We have effectively distanced ourselves from ourselves in order to survive.

But we can never be happy when that aspect of ourselves that was most damaged through others' neglect becomes further damaged by our own neglect. Healing does not occur through perpetuating the disease. That which is caused by neglect is unlikely to be healed through neglect. Rather, it is healed through attention—in this case, ours.

Letting Go of That Which Saved

By neglecting ourselves, we perpetuate the problem. We must reverse course. First, we must recognize that much of our distress is caused by the distressed part of ourselves that we are neglecting. This can be a difficult task. We are rarely taught that our psychological self is comprised of many functioning parts, only one of which is our ego (that aspect of ourselves that is aware of ourselves). Nor are we taught that some aspects of ourselves are fixated at certain ages where the trauma occurred. Few of us are likely to recognize that there is a damaged child living within our personalities.

Consequently, simply recognizing the wounded child within is a formidable task. The second step is even more difficult. We must connect authentically with our wounded and abandoned child. While Jungian analysis provides many ways to do just that, far too few know about it. Even fewer enter analysis.

Nevertheless, there are options. One is active imagination. Another is dream analysis. For those of us for whom analysis is not an option, there is a third way. We can teach ourselves to conceive of those emotions that are clearly inconsistent with adult functioning as our neglected and abandoned children within. These emotions are encoded in synaptic configurations that, because they fire in the present, feel as if they refer to current events.

They don't. They are flashbacks that reconstitute in the present the deeply traumatic experiences of the past. The intensity of our emotions is related to the past trauma, not to the present situation. Recognizing that reality liberates us from thinking that there is something wrong with us as adults. As adults we are fine. We function well and react to situations appropriately.

But our inner traumatized children do not. Their behavior is completely disproportionate to present circumstances. Most importantly, just as they needed secure and loving adults to help them to cope with challenges in childhood, now they need us to fill in that gap. What they needed then is what we are now.

Oneself As Parent

Remember morphogenesis? The psyche is always trying to increase its ability to survive. Our job as analysts is to facilitate that process in our patients. We are not the process. We are simply the catalysts. How do we do it? We'll look at that in Part Two.

For now, thank you for wading through all that has gone before in the name of preparation. We are now ready to enter the clinic. This is not some cloistered-off space on the margins of society. Rather, it is itself a society, whose twin goals are prophylaxis and healing. Let's review what we have learned.

In Chapter One we saw how the basic mechanics of organic matter contain templates for all that our psyches strive to achieve. Biochemistry contains the means not only to meet challenges to the organism, but to obviate them as well. Think of parasites and antibodies. But, in line with the shadow side of everything, biochemistry is itself a challenge. Challenge and adaptation do not cancel each other out, however. They comprise a basic archetypal bifurcation that oscillates and compounds. Jung called them *tension of opposites*.

In Chapter Two we looked at how molecular dynamics contribute to psychological dynamics. Jung's structure and dynamics of the psyche, especially as explained in his theory of archetypes, provides an analog to much that we are discovering today about how matter regulates itself. To discount the role of physical chemistry in our personalities is to worship at the altar of the god of ignorance. Jung wrote that archetypes are the psychological components of instincts.[77] And instincts are functions of

[77] "Instinct and the Unconscious," *The Structure and Dynamics of the Psyche,* CW 8, par. 277.

physical chemistry. It is this transition from chemistry to psychology that most concerns us in this study.

In Chapter Three we compared Jung's theories with chaos theory. We discovered that tensions of opposites consist of expanding systems of bifurcations and oscillations, which frequently generate patterns, or *attractors*, that express their dynamics in images. Thus, chaos theory can do much to explain how individuation progresses. SDIC, self-similarity across scale, fractal dimension, mutual inhibition equations, the *r* factor—all help to clarify the complex dynamics through which organisms adapt themselves to their environments.

In Chapter Four we considered the paradox of defenses. We noted that Analytical Psychology locates the domain wherein physiology and psychology interact in the unconscious. Particularly before consciousness has developed sufficiently to perceive and analyze situations (thereby to choose behavior), the unconscious responds to environmental challenges on its own. These responses become encoded in synaptic configurations that can conjure threat when none is present. Thus defenses also threaten.

Finally, in this chapter we looked at trauma. Even after consciousness can make its own decisions, some events are so overwhelming as to bypass it altogether. Traumatic life experiences bypass consciousness and set up their own synaptic configurations. While these are always adaptive in the moment, they become less so as the likelihood of traumatic experiences diminishes. Then it is up to the ego to correct the situation.

We undertook these analyses to understand better how conscious and unconscious become alienated from each other. To put it anthropomorphically and metaphorically: Neither seems to understand the other. As we shall see, that can lead the two to work at cross-purposes.

Consequently, from here on we shall concern ourselves with healing the split between conscious and unconscious. We'll owe a debt to Jung's original theories and to the post-Jungians who have elaborated them. We've already seen what genetics, physical chemistry and microbiology can add.

We'll propose some basic methods of increasing our adaptation and furthering individuation through advancing relations between ego and unconscious. We'll discover how to reconnect with aspects of ourselves that may once have saved our lives but have since become liabilities. Rather than trying to put those aspects out of commission—for example, through medication or other forms of suppression—we'll learn how to embrace them in gratitude for all they once did for us. We'll apologize for leaving them behind as consciousness developed its ability to assume the tasks of adaptation. We'll integrate them into newer, more functional alliances to meet life's challenges.

In short, we'll discover how to recognize adaptations that became trapped in the synaptic configurations derived from previous challenges. We'll explore ways of connecting with them to liberate them from the past as we unite together into new, more mature and successful alliances to meet present and future challenges.

Through renewed relationship with the more troubling aspects of our psychological development, we'll finally come home to ourselves.

PART TWO
Into the Clinic

6
Clinical Paradoxes

> *That I feed the beggar, that I forgive an insult, that I love my enemy in the name of Christ—all these are undoubtedly great virtues. What I do unto the least of my brethren, that I do unto Christ. But what If I should discover that the least amongst them all, the poorest of all beggars, the most impudent of all offenders, yea the very fiend himself—that these are within me, and that I myself stand in need of the alms of my own kindness, that I myself am the enemy who must be loved—what then?*[78]

Defenses are paradoxical. To defend against something is to keep it in mind. Defenses empower that which threatens, regardless of the severity of the threat.

Rarely do we consciously construct psychological defenses. In childhood this is particularly the case. Children do not construct defenses, they just happen. In Chapter Four we saw how defenses oscillate between approach/avoidance, empowering/diluting, past/present. All of these are active in childhood defenses that persist into adulthood.

As we grow into adults, we become strong enough to survive much that was life threatening in childhood. Or, at least our ability to do so improves. Unfortunately, many of us remain trapped in relationships or jobs or political situations that render us as powerless as children. Adult development affords few options in implacable environments.

Our ability to survive or to escape adult traumatic situations notwithstanding, childhood defenses that were unconsciously put into place do not simply go away. They accompany us into adulthood. But we often fail to recognize them as remnants of childhood. In those situations where we are forced to admit that our control over ourselves is less than perfect, we tend to attribute that lapse either to momentary weakness or to an unreasonable environment. Rarely do we realize that the child we left behind still lives within us.

In this chapter we shall see how childhood defenses persist into adult-

---

[78] "Psychotherapists or the Clergy," *Psychology and Religion,* CW 11, par. 520.

hood. We'll propose a four-fold scaffold through which to identify them. We'll focus especially on how childhood coping mechanisms place us in paradoxical situations as adults when our defenses accomplish the opposite of what we intend.

This chapter teaches us how to identify defenses that originated in childhood. In the following chapters we'll see what we can do about it.

Adaptational Paradoxes

Before consciousness develops sufficiently to take the psyche's helm, the unconscious assumes the task. Physical chemistry, as implemented by genetics and synaptic plasticity, seems to have a mind of its own. It assesses situations, mobilizes appropriate responses, and records the results. The bifurcations and oscillations of deterministic chaos underlie many of these activities.

In the last chapter we looked at how the unconscious copes with life's challenges before consciousness develops, and how its actions become encoded in synaptic configurations. Later in life, when these configurations become active, we experience them as if they are responses to current events. Actually, they are reconstituting—in the present—previous reactions to previous events. Trauma theory contributes much about how those reactions originally became encoded synaptically.

Therefore, many of the more extreme responses that we experience in adulthood have less to do with current situations than with previous challenges. Essentially, they are flashbacks. We would do well to forego accusing ourselves of being irrational or disordered. Rather, we should see them as the reappearance in the present of unintegrated aspects of our personalities from the past.

Healing the split between unconscious emotion and conscious cognition makes it possible to reclaim our childhoods while reassuring ourselves that we are pretty well adjusted to adult life. With that we can come home to ourselves. We can realize that we are not crazy. More importantly, we can both relive and reconfigure the past.

To say that we cannot change the past is to imprison ourselves in the external reality of the present. The past is not over. It is preserved in synaptic configurations in our brains. By engaging those configurations we connect them synaptically with other areas of the brain. Rather than the synaptic residues of childhood trauma connecting directly to the amigdala, we can connect them with the cognitive capabilities of the frontal lobe.

Healing is connection. Healing is relationship. By bringing into relationship aspects of our personalities that were previously estranged, we promote wholeness, for wholeness is itself healing.

That much of our brain mechanics involve bifurcations and oscillations means that much of our psychology is paradoxical. We think we are doing one thing when actually we are doing another. Or, the very means by which we attempt to affect a certain result ensures exactly the opposite. That the various aspects of our psyches appear to be working at cross-purposes can make us feel that either we or those around us are simply nuts.

Neither is. We are simply caught up in a sterile oscillation that cannot bifurcate further. It's stuck—and so are we. In order to free it to develop new patterns of perception and behavior, we must bring it into consciousness, creating thereby a new bifurcation between conscious and unconscious that can oscillate further into new patterns of perception and behavior.[79]

Now we'll look at some examples of these paradoxes, specifically those where we cope with life's problems in ways that actually exacerbate them. For instance, consider a man who cultivates low self-esteem as a way of heading off humiliation. He believes that the closer he is to the ground (the *humus* in humiliation), the shorter he'll fall. Unfortunately, his low self-esteem leads others to look down on him, generating the very humiliation he fears. Feeling betrayed, he responds with anger. How could anyone disparage one so humble?

Or, in order to avoid disappointment, a woman refuses to ask for what she wants. So she never gets it. Likewise, by never asking questions that she fears have no answers, she lives in constant fear of unanswerable questions. Finally, by seeking to avoid attack by not speaking up, she gets steamrolled. Always, the defense she adopts against the outcome she

---

[79] For further elaboration of this process, see *Archetypes,* pp. 26ff.

fears produces that very outcome. Trying to avoid life's bifurcations—in her case between contentment and disappointment, complacency and uncertainty, security and attack, courage and fear—leaves her trapped in a deterministic chaos that keeps producing the same result.

Early childhood defenses always involve the persona. To get our needs met, we adapt ourselves to what we feel others need to see. This can be a lose–lose proposition. If others respond favorably to us, we may think that it is based on a false impression (the image of ourselves we have constructed to attract them). If they respond unfavorably (or ignore us), then we have failed to adapt successfully. As a result, we may be inclined to adapt even more intensely, which exacerbates the situation.

The persona serves to reveal that to which we hope the environment will respond favorably and to conceal that to which it might not. Thus, it reflects who we are only to a limited degree. Moreover, persona relationships are always hollow. We do not know if we are accepted for who we are or only for how we present ourselves.

Eventually, the shadow appears. That which is excluded from the persona begins to leak through. Then we are in real jeopardy. What will happen now? Will we be accepted or rejected? When this crisis invades an intimate relationship, the couple often decides to split up—or seek therapy. When it is a defense, a persona generates paradox.

Because persona development begins at birth, much is constructed before we realize it is happening. Parents always prefer that we present ourselves in certain ways that to some degree reflect who we are, but to others do not. That can present some very interesting dynamics in analysis on the transference/countertransference interface when we treat our analysts as we treated our parents—or wish we had treated them!

Consider a woman whose unconscious realized early in life that survival depended on dodging her family's negativity. So it developed a persona that would prompt people to idealize her. Now, later in life, whenever she believes that she has failed to live up to others' idealizations, she feels showered by negativity. The original dilemma lives on, sustained by the very means by which she tries to avoid it. Paradox.

We tend to assign fault to those whom we idealize when they fail to live up to our expectations. Rarely do we blame ourselves. Thus, this woman's persona actually invites attack by engaging idealizers who are prone to be attackers. In effect, what exists between her and others is the same dynamic that exists between her shadow (feelings of inadequacy) and her persona (competence and expertise). The result? Fear and bravado are stuck in a sterile oscillation.

Another common paradox is reflected in a man who fears deprivation. Wanting something he cannot have or losing something he values is so painful that he studiously avoids desiring anything or anyone. No wants, no disappointments. He keeps himself in a state of constant deprivation. Paradoxically, the very means by which he tries to protect himself from deprivation ensures that he is always deprived.

Here's another fairly common paradox. A young man always plays the victim in hopes of gaining others' assistance, sympathy or favor. Unfortunately that leaves others feeling victimized by his playing the victim. He feels victimized and so do those around him. His victimization is contagious. So he never gets what he wants. Even worse, others who feel victimized by his playing the victim often expect him to minister to them. His life is one long competition for victimhood.

Both of these defy logic. Which is precisely the point. They are not adult adaptations, but childhood defenses. They were constructed by an unconscious that realized that the environments in which they lived were so deprived that survival was compromised. So it arranged perceptions and behaviors that compensated for what was missing in the environment.

When situations arise in adulthood that recall the childhood environment (that is, recall the threat), adult adaptations are preempted by childhood adaptations. While we may realize that we are acting like a child, we rarely recognize that as the child within taking over. That is not surprising. Few of us are ever told that our psyches are comprised of a number of independently functioning centers. How could we know that the children we were are forever alive in our psyches?

Because they benefit from more highly evolved cognitive capacities, adult adaptations are usually more functional than childhood adaptations. The latter, however, are linked more directly to the emotional centers in the brain. Thus, they activate more rapidly and more powerfully than adult adaptations. Then we find ourselves in a paradox. We may know cognitively what works better in any given situation, but the childhood adaptations beat us to the punch. They are up and running before we have time to decide what we would like to do.

The paradox of child functioning in adult behavior is not the only paradox associated with defenses. Perhaps the most confusing is the manner in which defenses often reconstitute the original threats. That which should protect, threatens. That which should be the solution becomes the problem.

Let's look at some examples.

The Paradox Scaffold

Defense-generated paradoxes follow a pattern, which is itself chaotic. That is, responses to dilemmas create bifurcations and oscillations that preserve the original conflicts. We can construct a four-level scaffold to help us to see the essential dynamics of these paradoxical situations. For brevity, we'll code these levels as follows:

ft for the feeling tone(s) that permeate(s) the expressed problem.

def for the defenses through which we try to protect ourselves from feelings.

res for the result or outcome of employing the defenses.

pp for the paradox that defines the problem.

We'll start with Betty.[80]

Betty

Betty feels trapped in relationship. She feels like she is always under the other's power. While such situations are always very complex, we can simplify the feeling tone like this:

ft: trapped, under others' power

To manage these feelings and to protect herself from the situations that activate them, she defends by withholding herself so that no one can

---

[80] As with all examples in this book, names have been changed.

engage her. We can represent that defense like this:

def: withholds herself so no one can have a piece of her

Obviously, that lends itself to hypervigilance. She monitors the environment for any suggestion that someone might entice her into an entrapping relationship. Her life is essentially controlled by the other, whether or not the other is trying to control her. The result:

res: everyone from whom she withholds herself has power over her

The paradox of her defense is that her attempts to avoid being controlled by others gives others an enormous amount of control over her (again, whether or not they seek control). That constitutes the paradox:

pp: she is trapped by the defending against being entrapped

Betty's hypervigilance began in childhood. Her parents were very controlling. The defense (withholding herself) was the best that her unconscious could implement when her consciousness was insufficiently developed to determine a better solution. While that may have worked pretty well in childhood (and, in fact, it did), it doesn't work in adulthood. Rather, her fear of being controlled has become a self-fulfilling prophecy.

The bad news is that a defense that may once have been adaptive is now maladaptive. The good news is that the original problem is preserved in the paradox. She can now revisit her childhood and move things in another direction.

The solutions we think of as adults, however, do not always dilute the feeling tones that originated in childhood. Emotions are usually more powerful than cognitions.

The problem is deeper than control. When we parents overly control our children, we are not relating to them authentically. However noble our motives, we are treating them as objects to be manipulated. The key to the situation is the word *overly*. We have to control our children to keep them safe, to teach them social graces, and to actualize their desires for learning and development. Like Goldilocks, we must avoid too little or too much control in favor of the "just right."

To say that this is difficult is to wildly underestimate the challenge.

We do the best we can. The best solution to the dilemma of discerning what level of control is enough, too little, or too much lies in the chaos of the situation. If we can bring our oscillating chaos (our own doubts) into a bifurcation with theirs, the two oscillating systems can form a new attractor that can not only bring us together, but produce a better solution to the problem than either of us could imagine. That's a fancy way of saying that in many cases where control is an issue, it is best to collaborate with our children. Obviously that will take different forms at different ages.

Again, overly controlling parents do not relate authentically to their children. While control may seem to be the main problem, it isn't. The main problem is loneliness. Our parents do not see us as who we are, do not treat us as people, do not engage us in openness and love. Unfortunately, to find better alternatives as we age is to leave that lonely child behind. The child we were remains stuck in a lonely home. We escape into adult life, where we become the ones who try to control the child.

But leaving the child behind does not mean that he or she goes away. Rather, he or she sinks into the unconscious where he or she becomes active when-ever our adult situation mimics the childhood one. The ideal solution is to create an authentic relationship in adulthood with the lonely child who still lives within us and is stuck in that stage of development where being overly controlled became a problem. In short, we must be the good parent to ourselves.

The woman who presented this problematic scenario came into analysis to get rid of the feelings. But those feelings are who she was when she first became aware of the problem. To eliminate the feelings is to eliminate herself. Anything that serves that purpose—medication, therapy, cognitive restructuring, dream analysis, insight—constitutes self-rejection, which perpetuates the original problem.

Now that we have a scaffold to help us to understand these kinds of feelings, let's look at more examples to see if we can gain some insight into how we might heal ourselves more constructively. For example, what might we make of the following scaffold?

Lucy

ft: trapped by life, fear
def: does nothing, refuses to try anything, makes no moves, withdraws from the threat
res: paralysis
pp: by refusing to make any move, 1) is trapped by the defense, and 2) never finds out if she really would be trapped by what she does

Like Betty, Lucy also feels trapped. And, like Betty, she defends by avoiding anything that might entrap her. That leaves her essentially paralyzed. In fact, the defense itself now entraps her.

Once again, the defense preserves the original problem. Fear of potential entrapment reconstitutes her actual childhood experience. Lucy fears that doing anything will lead only to being more entrapped. Yet, doing nothing is also a trap. As an adult, she lives in the world she experienced as a child. She never gets over the fear and never discovers whether ignoring it and living life will either liberate her or entrap her.

The analytic container gave Lucy the security to transform her childhood dilemma by opening herself to her fear. She was then able to provide as an adult that which had been missing in childhood. It was as if a loving adult had entered her life when she was a child, and had explained to Lucy that it was not she, but the environment, that was the problem. The loving adult could comfort the child and be a strong companion. She could help the child to grow. She could accept her without trying to control her.

Lucy could relive and redo the past. What had formed in childhood had endured into adulthood. Recognizing, accepting, and holding her fearful child self could revise the defense that had originated back then. Her child self could grow up. Lucy would be less controlled by, less entrapped in, childhood emotions. The trap of the childhood defense could be transformed into a healthy adult relationship, first with herself and then with others in her world.

Bill

Bill was also afraid. His scaffold looks like this:

ft: fear of incurring wrath or blame
def: withdraws when in murky waters
res: feels a heaviness and futility and often feels criticized for withdrawing
pp: withdrawal incurs the very wrath it is meant to avoid and makes him feel he's to blame

By not entering fully into life, especially during difficult times, he authenticates his fear of being to blame for the outcome of situations. Like Betty and Lucy, his defense is one of paralysis in the face of his fear. Unfortunately, it leaves him open to charges that he is not acting responsibly when crises occur.

We can imagine what Bill's childhood must have been like. Did his parents blame him for doing or not doing what he was too young to understand? Was there no one to moderate their effect on him? Was he alone in his fear? Did his unconscious put together the best response it could? Does the resulting defense plague him to this day?

Pretty much. But again, like Betty and Lucy, he can re-work the past. What was missing in his life as a child exists in his life as an adult. And, like Betty and Lucy, it is he. Simply accepting his feelings begins the process of transformation. Then seeing his feelings as the child he was and continues to be gives him the opportunity to transform them by enabling the development he missed as a child.

Remember: To try to escape from or rid oneself of troublesome feelings that originated in childhood is to reject the child we were—and are. The loneliness and isolation we felt as children are magnified and amplified by rejection in adulthood. Now, we are the ones rejecting ourselves.

Which gives us the opportunity to provide for ourselves the healing we lacked as children. That is the beauty of analysis. Everything that ever existed in the unconscious exists throughout life. We can relive the past. We can redo what happened. We can repair the damage. All it takes is to understand that our most troublesome feelings are ourselves as children crying out to us from the past.

In analysis, Bill discovered that he could be the agent of his own transformation. No longer was he the victim of his own feelings. Now, he was the healer.

Alice

Alice came into analysis because she felt deficient and inadequate. After we looked more deeply into her feelings, we put together the following scaffold:

ft: deficient, inadequate
def: does everything for everyone else
res: exhaustion and feeling inadequate, a failure
pp: trying to be overly-competent leads to feeling inadequate

Unlike Betty, Lucy and Bill, Alice defends against her feelings of inadequacy by overly compensating. She does the opposite of the previous three. Unfortunately, no one can do everything. No one succeeds at everything. So she is constantly confronted with her own failure.

The defense reconstitutes the feeling tone. From the perspective of deterministic chaos, she is stuck in sterile oscillations that serve only to regenerate the original pattern. She needs something to raise the *r* factor. She needs to generate a new attractor. Until she does, everything is just stuck.

Mabel

Mabel has the same feelings, but approaches the problem somewhat differently.

ft: inadequate
def: tries to come across as totally competent
res: others do not share ideas and perspectives with her
pp: trying to appear competent robs her of others' input, thereby making her inadequate (not benefiting from others' insights, thereby never seeing how they may be wrong or inadequate)

Like Alice, Mabel tries to compensate for feelings of inadequacy by trying to do too much. Not only does that approach leave her feeling inadequate, for—like Alice—no one can do everything, but Mabel suffers a double blow: Trying to be totally competent leads others to avoid sharing ideas with her. She bears the burden of solving problems alone. No one helps. Either through passive-aggressive envy or simply feeling less competent, others avoid pitching in.

As in childhood, Mabel is left alone in her dilemma. No one helps her. If they did, her feelings of inadequacy would likely be magnified. In failure or success, she remains alone. Throughout, there is no one with whom to share the burden. Inadequacy and loneliness oscillate endlessly.

She enters analysis to figure out the problem so that she can get rid of her problematic feelings. Of course, we both wished for that. But before the feelings can abate, we need to accept and embrace them. That is often too paradoxical to make sense. But it's the only way. Only by entering into the paradox of our defenses can we transform the past, for by entering into the paradox we enter into the childhood dilemma that necessitated the defense.

The Scaffold in Analysis

Rarely does the scaffold appear fully formed in analysis. More often, it is pieced together over time as different aspects emerge. Analyst and patient are united in their effort to imagine how the rest of the scaffold looks when one of its dimensions emerges. That interest is critical to healing the wounds of childhood trauma. Firstly, that interest brings analyst and patient together, ameliorating thereby the patient's loneliness. Secondly, when we begin to see what afflicts us as an interesting puzzle to be engaged and figured out, despair is turned into hope, fear into excitement.

Let's look at some examples to see how it works. To replicate the analytic process, in the following scaffold the result of the defense and the paradoxical situation have been removed.

Gladys

ft: fear that intimacy will be harmful
def: scrutinizes others to see if there is anything to fear

Gladys is so afraid of intimacy that when it offers itself to her, she tests it *ad infinitum* to see if there is anything to fear. She insists that others reveal themselves totally to her. What might be the result of that?

Her intensity pushes others away. Nothing they do or say reduces her fear. Her ceaseless testing of their sincerity is terribly frustrating. As everything she does is designed to eliminate her fear, fear infuses everything she does, for trying to do away with a feeling establishes it as a point of reference. She simply cannot resist monitoring the environment for any evidence of its presence. As such hypervigilance puts others on edge, her feeling tone activates the same in others, like a sympathetic vibration.

Gladys's obsession with total transparency sabotages intimacy before it has a chance to form. Metaphorically, she drags the baby out of the womb before it is ready to be born. It's not that she doesn't want intimacy. She simply fears that intimacy will be hazardous. By frustrating intimacy before it can develop, Gladys never finds the intimacy she seeks. That's the paradox.

Her insistence on transparency and openness in the context of intimacy is too much for others to accomplish. It makes them feel inadequate and trapped, whereupon they flee. Thus, the *pp* is simply that her defense against being hurt in intimate relationships ends up hurting her by denying her the healthy intimacy she seeks. Intimacy requires trust. Her lack of trust breeds the same in others.

The defense does not accomplish what it is designed to do, namely, do away with the fear. Once again, that's the bad news. The good news is that she can now resonate with others. We all have some fear of intimacy. That can be a basis on which we connect with each other. Thus, Gladys's own fear provides a means to connect with that of others.

If only she can allow herself to do that. The first step is to accept her own fear. By entering more fully into relationship with her own feelings, she can then recognize and enter into relationship with those of others. She might even be able to share the original problem with others now that she understands the defense and its result.

Assembling the Scaffold

As we said, very rarely do all four elements of the scaffold appear early in analysis. But one or two are usually present. It is our challenge to figure out what the others are. For example, consider the following result—

humiliation, frustration and alienation—that accompanied feelings of low self-worth. Both appeared early in the analysis.

Albert

ft: lack of self-worth
res: humiliation, frustration, and alienation

Albert was aware of the feelings (humiliation, frustration, and alienation) but could not link them to any defense against something else. As we looked into his home and employment situations, we discovered that he relied on his wife for self-esteem. In fact, as we delved more deeply, we discovered that he relied pretty much totally on his wife for affirmation and confirmation of his self-worth.

That seemed to qualify as a defense. So we added to the scaffold:

def: reliance on wife for self-esteem

Why would he do that? Did he suffer from chronic feelings of low self-worth? That seemed unlikely, as he was a very successful businessman who was lionized in the local media for his accomplishments. However, accomplishments—particularly the kinds of accomplishments that are based on over-achieving—often compensate for low self-esteem. That one succeeds doesn't necessarily mean one feels successful. Sometimes, one feels it's all a sham. That recalled the result:

res: humiliation, frustration and alienation

That turned out to be the case. Was it also the paradox (low self-worth in the face of enviable success)? While that is certainly a paradox, it turned out not to be the one that gave rise to the dilemma. For that, we had only to look at the feeling tone, which was "lack of self-worth." In other words, the feeling tone gave rise to the defense (total reliance on his wife for self-esteem). Why was the result (humiliation, frustration, and alienation) not as intended (comfortable levels of self-worth and self-esteem)?

As we explored further it occurred to us that looking to others for self-worth might itself undermine self-worth. We proposed that as the paradox:

pp: looking to others for self-worth undermines self-worth

That made sense: To rely on another to confirm our self-worth is to undermine intrinsic feelings of self-worth. Our self-esteem is conditional, not on our own estimates of ourselves, but on others'. Looking to others for confirmation of our worth carries with it impressions of dependency, weakness, uncertainty, lack of core values, shallowness. What we do to curry those expressions of esteem can lead us to feel obsequious, sycophantic, pitiful, even pathetic. What we gain—if anything—feels like pandering.

So there we had it. The scaffold in this case looked like this:

ft: lack of self-worth
def: looks to others (particularly his wife) for affirmation and confirmation of self-worth
res: humiliation, frustration, alienation
pp: looking to others for self-worth undermines self-worth

Fair enough. What, then, is the solution? We'll get to that in the next chapter. Meanwhile, here's a hint: It's the same as the examples cited above. It involves discerning the origin of the feelings of low self-esteem and self-worth. The dilemma consists of an ongoing reenactment of what in childhood was a highly successful adaptation to a highly negative and undermining environment. Problem is, what succeeds in childhood often fails in adulthood. Adults are not supposed to behave like children.

Bruce

Sometimes the paradox is pretty obvious. Bruce presented with feelings of frustration. While these focused primarily on people and events in his daily life, none seemed extreme enough to warrant the intensity of his feelings. Something else must have been at work. Perhaps his frustration was not the core feeling, but a result of how he tried to cope with his core feeling.

In fact, that was the case. Underlying his frustration was anger. Extreme anger. Murderous rage, actually. While the rage rarely appeared, frustration was almost constant.

Why was that? As we looked more deeply into his frustration and anger, we discovered that he coped with his rage by trying to understand what set it off. He would then adjust his response to something less or other than rage. Foremost among these lesses and others was frustration. So far, the scaffold looks like this:

ft: murderous rage
def: understand and master the rage
res: frustration

That leaves us with a bit of a dilemma. How could understanding and mastering rage possibly be a problem? Moreover, where's the paradox in that?

After thinking about it for a while, the answer became clear. His attempt to understand and to master his rage was directed at himself. He did to his rage what he feared it would do to others. But the rage didn't abate. The only thing that changed was the object of his rage. He was transferring his rage from others to himself.

Bruce was doing to himself exactly what he was trying not to do to others. He was trying to murder his murderous rage. There's the paradox. His murderous rage, which served to suppress others, ended up being suppressed by him. He did to himself what he feared his rage would do to others—murder. As in the previous example, others felt his rage in every one of his attempts to suppress it. He created the very situation he was trying to finesse. Consequently, our scaffold now looks like this:

ft: murderous rage
def: understand and master the rage
res: frustration
pp: by trying to understand and master his rage, he tries to murder it, doing to himself exactly what he's trying not to do to others

Relationship

As we saw earlier in Chapter One, relationship is the essence of morphogenesis. Deterministic chaos, which generates patterns that provide new opportunities for development, is the result of bifurcations that oscillate with one another. As bifurcations and oscillations compound, complexity arises in the system. That complexity is the essence of relationship (which, as the first sentence stated, is the essence of morphogenesis).

Consequently, problems in relationship within ourselves and between ourselves and others frequently, if not always, indicate frustrated morphogenesis. Why does it stall? Because we often prefer the comforting familiarity of the *status quo* to the intimidating uncertainty of change. Thus, problems in relationship serve to restore healthy relational dynamics.

A common complaint is that we feel like we have failed (or are failing) in relationship. Sometimes our significant other informs us of this and sometimes we discover it on our own. Either way, we feel confused. How can relationship be failing when we are doing what we have always done and that worked in the past? Well, as we just saw, that's exactly the problem. Let's look at some typical variations on the theme of relational failure.

Adriel

Adriel's feeling tone was curiously nonspecific. She feared that she was a failure in relationship. As with most feeling tones, however, it didn't specify itself in any particular aspect of relationship. It was simply an overall feeling—an affective ambiance, so to speak.

Our search for the origin of the feeling did not result in much. We knew that she feared that her relationships were a failure. Just what was activating that fear, however, wasn't revealing itself.

Or was it? However contrived and truncated, the analytic relationship is a relationship. Would Adriel's dilemma take up residence there. What was occurring on the transference–countertransference interface?

At first, nothing seemed out of the ordinary. But then, several things began to happen. Firstly, she would become very upset if she arrived late for the appointment. Secondly, as the end of the session approached she would become anxious about not going over the time. Thirdly, if she hadn't copied—or even written in—her journal, she would be extremely apologetic. What was going on? She seemed to be afraid to commit errors or to be in any way inadequate. Was she trying to be perfect? If so, was that the defense?

As we looked more deeply into that possibility, it seemed to be the case. She began to see that she defended against her fear of failing in relationship by trying to be perfect. She tried to be the perfect wife, mother, daughter, friend, patient. Whatever she perceived others to expect of her, she strove to accomplish.

She was surprisingly good at it. In school she had maintained a very high grade point average. Unfortunately, she never made it to the top. She missed being valedictorian by a few hundredths of a point. In college she was active in sports and still managed to achieve a high average. But she was never the top in either athletics or academics. Able to reach the heights of the stratosphere, she never made it into orbit.

We had a clue to her fear. No one can be perfect and complete. The two are opposites. "One trick pony" describes perfection. "Jack of all trades, master of none" describes completion. Adriel was trying to be perfect in everything. That, of course, was impossible. So the result of her defense was exactly what she feared: That she was a failure in relationship.

Thus the paradox: To be completely perfect is impossible. Even if it is achieved to a high degree in some areas, the amount of time and effort it requires leaves us deficient in others. Failure looms. Hopelessness becomes despair. It all just feels awful.

We proposed the following scaffold:

> *ft:* fears she's a failure in relationship
> *def:* tries to be perfect
> *res:* feels she's a failure in relationship
> *pp:* because perfection is unattainable it creates the very feelings of inadequacy it's supposed to offset

Perfection is frustrating in itself. In relationship, however, perfection is a curse. Alfred Adler did us a great service in articulating that fact in his school of Individual Psychology. There he revealed that what we feel most connected to in each other is our inferiority.

And Jung:

> A human relationship is not based on differentiation and perfection, for these only emphasize the differences or call forth the exact opposite; it is based, rather, on imperfection, on what is weak, helpless and in need of support—the very ground and motive for dependence. The perfect have no need of others, but weakness has, for it seeks support and does not confront its partner with anything that might force him [sic] into an inferior

position and even humiliate him [sic].[81]

Few—if any—can identify with the perfect. The imperfect, however, is familiar and comfortable. We feel little threat in the imperfect. We feel no need to compete with it. We can embrace it.

We decided that she needed to try to be less perfect. If she could do that (while not trying to do it perfectly), she could offer others the opportunity to connect with her as equals. But that opened up another problem. What if she was afraid of relationship? We are all to some degree fearful of relationship.[82] In the extreme, however, difficult challenges ensue.

Relationship thus became the deeper focus of our work together. That, in turn, revealed another scaffold underlying this one. That scaffolds illustrate defenses leads them to be fractal, that is, scaffolds upon scaffolds. In typical deterministically chaotic fashion, analysis reveals scaffolds of scaffolds.

Gary

Gary sought analysis because he consistently found himself embroiled in negative relationships. He could not figure out why. Even when he realized a relationship was negative, he persevered. Actually, he would immerse himself more deeply. That didn't make sense. Like Betty and Lucy, he felt trapped.

In view of that, we might surmise that he is a masochist. Maybe he likes being abused. Perhaps for him love and pain are synonymous. Did he have an abusive childhood? Did he learn that negative relationship was authentic relationship? Was negative relationship the only kind he experienced?

Questions about his past may be helpful in ascertaining why he feels the way he does in the present, but too often they turn out to be dry holes. All of our drilling turns up only the same thing at deeper levels.[83] Deterministic chaos and archetypal dynamics just keep turning out the same patterns. It might be better to delve deeper into the phenomenon at

---

[81] "The Undiscovered Self," *Civilization in Transition,* CW 10, par. 579.

[82] See "intimacy," above.

[83] Scaffolds of scaffolds can be hopelessly confusing.

hand. In this case, that's exactly what we did.

Gary had an idea about why he held onto, even embraced, negative relationships: He felt that he could transform them into something better. Perhaps by throwing himself into them and giving it all he had, something good would come of it.

That gave us something to work on. Did he have a martyr complex? Was he the type who roamed the world looking for messes to clean up? Was he an altruist who sought to make the world a better place? Was he of a particular religious bent that led him to sacrifice himself for others?

None of those seemed to fit. He didn't see himself as either a saint or a savior. Actually, relationships frightened him. It baffled him that he sought them out and, when he found them, dived more deeply into them the worse they were.

That was also helpful. Staying in negative relationships could be a defense. If so, against what? The answer was not difficult to find. If he was embracing the other in order to transform the relationship into something better, it seemed reasonable that something was driving him to do that. What might that be?

As we are seeing repeatedly in these examples, it turned out to be fear. He was so afraid of the negativity that inevitably surfaces in relationship that it (negativity) became the prime focus of all his fears. Learning how to deal with it constructively felt way too precarious. He preferred to obviate it.

Gary's fear of negativity was becoming the defining element in relationships. Rather than negativity being a subset of relationship, relationship was a subset of negativity. His fear of being destroyed by the negativity of relationship had become a self-fulfilling prophecy.

That was the feeling tone: Fear that negative relationship would overwhelm him and ultimately lead to his demise. That clarified his defense: His efforts to embrace the negativity reassured him that he continued to exist. He defended against being negated by affirming the relationship. There was the paradox.

The result was relatively easy to figure out. Embracing the other in a negative relationship kept him in the negative relationship. He was trapped by his defense against his fear. That led to the ultimate paradox:

Trying to alleviate his fear of being negated in negative relationships kept him trapped in negative relationships. The solution had become the problem. The cure had become the disease.

Here's the scaffold:

ft: fear of being killed/overwhelmed in a negative relationship
def: embraces the other to try to transform the relationship into something better
res: stays in the problem
pp: by embracing negative relationships, stays caught in them

Reprise

We might take a moment here to recall what the scaffolds tell us. As we have seen, bifurcations and oscillations are the hallmarks of morphogenesis. In order for development to proceed, oscillating dynamics must bifurcate further into new oscillations. We call this deterministic chaos. The constantly elaborating bifurcations and oscillations of deterministic chaos make it possible for more complex dynamics to form.

However, unless the deterministically chaotic system bifurcates with another system, the same patterns continue to emerge. In this way archetypal patterns (fractal attractors) that form in childhood resurface from time to time as complexity compounds.

When a dysfunctional pattern is established early in life, deterministic chaos preserves it and re-expresses it in future dynamic elaborations. This offers us the opportunity to revisit earlier times in our lives to rework the troublesome dynamics. In other words, deterministic chaos allows us to redo the past. We can move back and forth in time in psychological space.

Unfortunately, today's views of reality and our place therein do not admit of the possibility of reliving the past. "You can never go back," "What's done is done," "You only get one chance in life," and other clichés reinforce that view. However, while such may be the case for historical time, psychological time has no such restrictions.[84] There, the

[84] This is the difference between synchronic and diachronic time; see *Archetypes*, p. 26f.

past is present and the present is past. As we saw in Chapter Two, experiences leave their traces in synaptic configurations. When those configurations become active, they present themselves as *now,* not as *then.* By bringing the past into the present, they make it possible for us to interact in the present with the past as the present.

"For us to interact in the present with the past as the present" brings up the most important bifurcation of psychological morphogenesis, namely, that between conscious and unconscious. Here is where psychological development achieves its greatest potential. For the two to oscillate with each other is the very reason consciousness exists. Unfortunately, those oscillations are discouraged from bifurcating when the unconscious is dragged into consciousness or vice versa. The former is the imperialism of cognition, the latter of psychosis.

The paradoxes that the scaffolds reveal are essentially stuck. If we do not bifurcate with other bifurcating and oscillating systems, our deterministic chaos cannot elaborate into newer and healthier elaborations of oscillating dynamics. It gets stuck in a rut. The defenses mobilized to mitigate the feelings wind up preserving the original conflict. Once that is known, however, we gain the possibility of elaborating the paradox into a conscious–unconscious dialogue. Just how we do that will be the subject of our next chapter.

Sterile Oscillations

For now, let's continue to see how paradoxes reveal conflicts that simply oscillate within themselves, rather than elaborating into more sophisticated adaptations. Let's go on to one of the most problematic conflicts of relationship: abandonment.

Like trauma, everyone has experienced some form of abandonment. From the most severe (loss of parents when young, for example) to the less so (rejection by classmates), to the least (momentary loss of contact with parents in a department store), we have all experienced abandonment. Consequently, the ways of coping with it are as numerous as there are people.

Gustav

Gustav presented with a paradox. He was always going the extra mile in relationships. He shared all he had and sacrificed his dreams and desires to the other. No matter how they responded, however, he felt abandoned. Whether they took advantage of him or matched his efforts with sharing and sacrifices of their own, nothing helped. He wound up feeling abandoned in every relationship.

At first we wondered if he was suffering from a chronic feeling of abandonment. It certainly looked that way. No matter what he did, it stubbornly persisted. But there really is no such thing in psychology as an isolated feeling. While a particular feeling may persist throughout different experiences, it doesn't arise from nothing. Something must persist along with it: An adaptation, an assumption, a consistent element within diverse experiences, or even an attitude. Even in the case of PTSD, some element in new situations sufficiently resembles an aspect of the original trauma to activate the feelings that arose therein.

Underlying Gustav's chronic feelings of abandonment had to be something else: a traumatic experience, perhaps. Unfortunately, he couldn't remember any relational trauma save the feelings of abandonment themselves. Nor was there anything in the family's history to suggest a traumatic incident had occurred.

As is often the case (as with Mabel, above), clues show up on the transference–countertransference interface[85] Not surprisingly, abandonment was foremost in his mind. Would I be there when he arrived for his appointment? Would I suddenly announce that the analysis was unsuccessful and that I was terminating the relationship? Was he boring me? How did he compare with other patients? What if his insurance benefits

---

[85] *Transference* and *countertransference* refer to relational dynamics learned in previous relationships that become transferred into current relationships. In analysis, *transference* refers to how patients perceive the analytic relationship and *countertransference* to how the analyst sees it. While the perceptions of both may be accurate and have some basis in the character of the analytic relationship, transference and countertransference elements are so powerful that they can force the analytic relationship into their molds.

ran out? If I had to cancel an appointment, was I okay? Had I become ill? Was I thinking of moving to another city? Would I raise my rates above what he could pay or insurance would reimburse? *What if* resonated endlessly with *What if?*

The feeling tone was palpable. Feelings of abandonment were not connected to actual situations. They preceded events. There had been no abandonments in our relationship. Rather, he was terrified that they would occur. Once again, the feeling tone that powered the paradox was *fear* of abandonment.

Admittedly, that should have been obvious from the start. But in analysis it is painfully possible for the most obvious of observations to escape the notice of either analyst or patient. We had been concentrating so diligently on the degree to which abandonment was present in his various relationships that we missed the all-pervasive feeling tone of fear that infused those relationships.

Now we could construct the feeling tone and the defense:

ft: fear of abandonment
def: surrenders self for relationship with others—shares everything

In order to prevent the feared outcome (abandonment), Gustav surrendered himself completely to the other. He would monitor—sometimes hallucinate—what he presumed to be the other's expectation. He would then strive to live up to it as fully as he could whether it was good for him or not.

The result was not what he expected. He had hoped that this would curry the favor of the other. Sometimes it would, sometimes it wouldn't. But behind the actual outcomes of his efforts lay a subtler one: By surrendering himself totally to relationship, he was violating himself. He wasn't paying attention to his own needs.

There was the result:

res: violates self, sacrifices self for other

As we shall note in the next chapter, he was ignoring the needs of his inner child, who had felt abandoned in childhood and was continuing to be abandoned in the present—by him. By continually sacrificing himself for the other, he was abandoning himself.

Now the paradox became clearer:

pp: by sharing with and sacrificing to others, he abandons himself, which leaves him feeling abandoned

His attempt to prevent abandonment by abandoning himself created the very result he feared. It really didn't matter how the other in relationship treated him. Even the most attentive and loving companion could not obviate his feelings of abandonment, for he had already abandoned himself. We cannot defend against feelings of abandonment through abandoning ourselves. Gustav reconstituted his childhood experience in his relationship with himself. In true chaotic fashion, he was stuck in the original oscillation between abandonment and feelings of abandonment. There was no new bifurcation to get the dynamic moving again. The pattern was fractal.

Thus, our scaffold looks like this:

ft: fear of abandonment
def: surrenders self for relationship with others—shares everything
res: violates self, sacrifices self for other
pp: by sharing with and sacrificing to others, he abandons himself, which leaves him feeling abandoned

As mentioned before, rarely is there only one scaffold. The complexity of analysis originates in the multiplicity of challenges that foster defenses, most of which evolve into paradoxes. For example, Gustav also had profound feelings of being "out of the loop." He felt that life took no notice of him, that everything was just out of reach. The theme of his life recalled the title of Milan Kundera's novel *Life is Elsewhere.*

No matter what he did, he could not shake these feelings. They alienated him from life, others and himself. He was miserable. At first glance, this looked like a variant of his abandonment feelings. When he felt abandoned by others, he also felt abandoned by life.

But there were nuances that suggested greater complexity. His feelings were less abandonment than inadequacy. While life did feel just out of reach, he wondered if he simply couldn't reach far enough. Maybe the problem was himself. It is certainly true that our dilemmas are often complicated by elements of our own making. Maybe he couldn't reach far enough to grasp what life had to offer. But the picture wasn't complete enough for us to know what to do. We needed more to go on.

It wasn't long in coming. Gustav soon realized that he believed that to give himself the courage to strive for what life could offer him, he had to be strong. He had to ignore his feelings and strive to do better. This turned out to be his primary defense.

His primary defense required him to distance himself from the problematic feelings. So he pushed them away, tried to escape them, minimized them—anything to undermine their power, which he felt would only pull him down into greater despair. In short, he put his feelings "at arm's length."

The result? He felt worse. Not only were his feelings (that he couldn't grasp what life had to offer) a problem, but his solution wasn't working either. He was beginning to feel like a total failure. Attempts to reassure him by pointing out that he was a success in seeking treatment or by showing him Michael Jordan's video on failure were to no avail.[86] There was no solution-focused answer to his dilemma.

His life began to improve when he grasped the paradox of his defense. By trying to distance himself from his feelings, he was compounding how distant he felt from life itself. By disrupting their connection to him, he was disrupting his connection with life. Everything for him was elsewhere. Deterministic chaos was locked in a system for which there was no external bifurcation, thereby preventing his psyche from initiating new ways for his psychic energy to progress. Again, we'll have more to say about this in the next chapter.

For now, here is Gustav's other scaffold:

ft: life is elsewhere, just not there for him, no potential, everything out of reach

def: tries to escape the feelings, puts them "at arm's length"

res: feels worse

pp: by trying to escape feelings, prevents them from connecting with him, thereby reinforcing the feelings (of everything being just out of reach)

---

[86] Michael Jordan has recorded a wonderful video on the relationship between success and failure. It can be viewed on YouTube.

This scaffold complicated and amplified the previous one. Both paradoxes can be transcended—loosened up, so to speak—if he would reverse the course of the defense. In the previous example, his defense was to abandon his feelings of abandonment by rushing headlong into relationship. Better to hold his feelings and allow them to express themselves. By accepting them, he accepts himself.

A superior solution to his concern that life is elsewhere would be to accept his feelings of not being able to connect with life. To connect with his feelings is to connect with himself. Then he can discover that life is not elsewhere. It is connected to him in his connection with himself. He can connect with the life that lives within him.

Fear

All these examples resonate with fear. Each person is afraid that if they do not employ her or his defenses, bad consequences will occur. Here is the link with trauma. Actually, the feeling tones are a form of hypervigilance. The paradox of the defenses is that they replicate the original problematic situations that necessitated the defenses in the first place. Thus, the paradox is essentially a stuck bifurcation that oscillates but fails to bifurcate further. There is no possibility of moving beyond the defense. All is flashback.

Remember: Children's defenses are not a product of conscious decisions. While children may be painfully aware of threats in the environment, that awareness is rarely a result of conscious analysis of the situation. Rather, it is a spontaneous feeling, one that can lead to a great deal of confusion.

Children do not decide to create a defense. Rather, it is done for them by the unconscious. Consequently, they often are not even aware that a defense has been deployed. This can make them prone to behaviors in adulthood that they can neither explain nor understand. Everything just runs on automatic. The conscious adult is left to try to figure it out.

Feeling tones endure. We may not understand why, let alone what to do about them, but they permeate our perceptions and drive our behavior.

Ruby

Ruby entered analysis with an autoimmune disorder that had become life-threatening. Her mother, fearing that her daughter could die, advised her not to do anything that might exacerbate the illness.[87]

Unfortunately, there was little in the mother's mind that did not constitute risky behavior. The daughter followed her mother's warnings with the result that she never participated fully in life. She didn't date, travel, engage in sports, drink, dance—virtually everything was proscribed. The result? The daughter was engaged in a kind of living death. That was the paradox: By trying to avoid death, she was essentially living in a death-like state. She was practically a zombie.

The scaffold was fairly easy to assemble:

ft: mother feared her child (Ruby) would die
def: mother told her not to do anything so she wouldn't get sick and die
res: a kind of "living death" via not participating fully in life
pp: by letting death be the point of reference, steps taken to prevent it kill off life's potential

Death had become a point of reference for life. Steps were taken that killed off life's potential.

In this case, the mother's fears precipitated a defense on the part of the child. The mother's fears dictated her daughter's life.

Who could blame her? What parent would not take whatever steps are necessary to preserve a child's life? This is not a situation that any parent would choose. But the situation is much more complicated than simply trying to avoid death. In this case, the defense is the threat. Such is also the case in the next example.

Ellen

Ellen's presenting problem was that she could not "get going in life." She was not married and had no prospects for so doing. She was employed as a nurse but found the work uninteresting and a chore. Her life

[87] Ruby's analytic experience is covered at greater length in Chapter Eight.

was a drag.

As we reviewed her family history she mentioned that her mother had always been very fearful. Something was always wrong in her life. In fact, Ellen had recently learned that her mother had struggled with depression during much of Ellen's childhood.

What might have been the consequences? As we looked more deeply into her relationship with her mother we realized that her mother desperately needed Ellen to be constantly happy. Somehow, Ellen's happiness ensured the mother's happiness.

That is not unusual. Our children's happiness greatly determines our own. Nevertheless, as we saw earlier, "Disorder is a matter of degree."[88] To make our children's lives contingent on what makes us happy is to rob them of their individuality. A most interesting example is when we vow to give our children all the advantages we never had.

Again, what could be wrong with that? Should we not as parents try to provide the best for our children? Of course. Things go awry, however, when we prod our children into life courses that we would like to have pursued, but that they either would prefer not to pursue or are completely unsuited for. I may have always wanted to be an astronaut, but that doesn't mean my daughter should be one. Showering her with physics may deprive her of a fabulous career in real estate.

Such was Ellen's case. She began to realize that the life to which she aspired (marriage, family, etc.) had little to do with what she really wanted. Her dilemma was simple: Sacrifice her happiness for her mother's vision of happiness, or pursue her own definition of happiness at the expense of her mother's. In that state of affairs, only one of them could be happy. Which would it be?

The paradox was simple: Neither was happy. The daughter was pursuing a lifestyle that did not agree with her. Consequently, it remained beyond her reach. The mother realized that Ellen was unhappy. Thus, she was unhappy, as well. The origin of the paradox was in the mother, yet the daughter shared in the suffering it caused.

Metaphorically, it seemed that Ellen's mother feared that her (the

---

[88] Sullivan, *The Interpersonal Theory of Psychiatry,* p. 305.

mother's) descent into Hades (depression) would take her daughter with her. She also feared that her subsequent ascent (the depression lifting) was not shared by her daughter. Distraught that Ellen had been left behind, she clung desperately to her for reassurance that she was happy. Thus she sought reassurance through Ellen getting married, having a family, etc.

Tragically, that made Ellen want to avoid her mother. In fact, Ellen was confined to her own Hades, not because her mother had left her there, but because her mother needed her to leave Hades so the mother would not slip back in. By clinging, the mother was losing her daughter. By clinging, she was abandoning her.

Eventually Ellen decided that only her happiness could restore her mother's. She would have to depart from her mother's needs and pursue her own. Which she did. She went to medical school, became a gerontologist, and lives a very happy life free from marriage and family. She regularly vacations with her parents and siblings. Paradoxically, Ellen's mother regained her daughter when Ellen liberated herself from her mother.

That is an example of how one person's paradox infects another. In the next—our last—example, we see the same thing. In this case, I myself was the owner of the paradox, which I enforced on my patients.

The Double Bind

One of the more problematic paradoxes is the *double bind*, in which we are caught in a "damned if we do, damned if we don't" dilemma. In my practice, this occurs whenever I give a dual message that contradicts itself. For example, I become uncomfortable when patients are angry with me. I was supposed to have integrated this fear long ago in my training analysis. Unfortunately, integrating feelings does not make them go away. In fact, it usually makes us more aware of them, and thus more sensitive to them.

Nevertheless, I let my patients know that they are free to express their anger in sessions. That works pretty well as long as their anger is not directed at me. When it is, I find myself trying to duck it. Or, I manage to arrange sessions so that anger never appears in the first place.

Patients usually sense their analyst's fear, however. If they get angry with me, I go into a defensive stance. Then they are in a double bind: In order for the analysis to succeed, they need to be able to recognize and express their anger. But if they do so, they fear that they might alienate or harm their analyst. At that point, neither suppressing nor expressing anger is an option. They are in a double bind.

What are they to do? Being in a double bind generally makes us angry. Now they are angry about being simultaneously encouraged and discouraged from expressing anger. Anger simply builds. Thus, my scaffold looks like this:

ft: my fear of patients' anger

def: avoid it if possible, duck it if not

res: patients are in a double bind: I tell them not to be afraid of their anger while my fear leads them to suppress it

pp: putting patients in double binds makes them angry. Encouraging them to overcome their fear of anger when I clearly do not want to endure it inevitably angers them, which I then try to avoid by accommodating them, or by telling them how good it is that they are expressing anger, or by one of several subtle ways through which I try to manage sessions.

Environments have a great deal of influence over us. In these last two examples, paradox extends from one person to another. In the first, from mother to daughter. In the second, from me to my patients. My ambivalence about anger generates paradoxical ambiguity in the analytic relationship. My patients pick up on this, become frustrated, want to express the anger I am encouraging them to express, yet know that if they do there may be negative consequences.

Conclusion

That is how we can diagnose childhood trauma. Now let's look at what to do about it. Please note that the psyche is extremely complex. Consequently, so is analysis. While paradoxes characterize oscillating systems that cannot bifurcate into greater complexity, that is not all there is to individuation or analysis. We are touching on only one aspect of both. Paradoxes may be part of every analysis, but they are not the whole of it.

Yet, understanding the relationship between traumatic experiences and stalled individuation is critical to success. In the next chapter we'll consider how to get things moving again. In the last chapter of *Archetypes* I stated the solution as "make it fractal."[89] That is what we are going to consider. Just how do we "make it fractal?" We'll discover that it's simpler than it seems.

We'll begin with an example that illustrates the kind of paradox that stymies individuation. We'll see that releasing the energy bound up in an oscillating system that cannot bifurcate further can lead to successful individuation and a happier life.

In the next chapter, we'll learn how.

---

[89] *Archetypes,* p. 155.

7
Reconnecting

The important thing is not to interpret and understand the fantasies, but primarily to experience them.[90]

We've looked at several examples of how deterministic chaos can stymie development by becoming stuck in repetitive fractal elaborations of the original conflict. We've also seen how the paradox of defenses both sustains that stasis and provides clues about how to transform it. Remember, like the physical chemistry of the brain, it's all a function of bifurcations and oscillations.

Here's another paradox: The following descriptions of conscious engagement of unconscious (including active imagination) constitute conscious appropriation of the conscious–unconscious dialogue. It's impossible to describe accurately, let alone to explain, what happens when conscious surrenders itself to unconscious. Consequently, the following descriptions are a bit of a contradiction in terms, for consciously describing unconscious dynamics brings everything into the realm of consciousness. That, alas, leaves out more than it includes.

Nevertheless, the conscious/unconscious bifurcation is the main factor in healing childhood trauma. As we have seen, childhood conflicts are encoded synaptically and become activated throughout life whenever circumstances recall the original conflict—flashbacks, in other words. Thusly do we find ourselves as adults in the curious position of perceiving and behaving as children. The problem is that an oscillating system (between childhood and adulthood) has become stuck; it cannot bifurcate further because conscious and unconscious have become split off from each other.

Our task is to bring conscious and unconscious into relationship with each other. This does not mean merging them into an amorphous blob, however. Rather, it involves establishing a dialogue that elaborates their

[90] "The Technique of Differentiation Between the Ego and the Figures of the Unconscious," *Two Essays on Analytical Psychology,* CW 7, par. 342.

oscillations through further bifurcations.

What does that mean? Simply put, it means bringing unconsciousness into consciousness, and consciousness into unconsciousness. The former is pretty much the way things are done in psychology, if not in Western culture in general. The latter is a bit more complex. Yet that is the unique contribution of Analytical Psychology to psychotherapy. Individuation proceeds on a two-way street. To go in only one direction is to deprive ourselves of healing and development.

Eleanor

For example, consider a woman plagued by attractions to older men who are unavailable. When they decline to choose her as a mate, she suffers a serious blow to her self-esteem. Equally disturbing are older men with whom she seeks a platonic relationship whose interests are solely sexual. She compensates by spending time on internet dating sites, but to no avail.

She's miserable. When older men reject her, she berates herself for being unable to attract a mate. As a result, she pursues older men all the more ardently. When they come onto her sexually, she feels like a whore. The enduring power of these patterns suggests that they are defenses against the feeling tone of a childhood trauma.

One might say that she suffers from a father complex. Fair enough, but of what sort? What were its origins? How and why is it active now? As we worked together with an eye toward these questions, several things began to emerge.

Firstly, when she was young she wanted a strong relationship with her father. Of that, unfortunately, he was incapable. She was left with wanting him but not being able to have him. As a result, she suffered with unrequited love. As she said: "It's true. I remember those feelings. I remember specific instances." Once again, naming something helps to bring to consciousness at least a bit of what is swirling around in the unconscious. This is the first stage of the work: Making the unconscious conscious.

Constant feelings of unrequited love depressed her and made her life unbearable. Not surprisingly, she blamed herself. As a child, she did all she could to make herself attractive to her father. But eventually her loneliness led her to isolate herself. She would withdraw to her room for hours on end, whiling away the time by—among other things—making dolls' clothes.[91] The futility of engaging in relationship led her simply to give up.

Secondly, as she grew older, school expanded her world, where relationships with peers, activities, challenges, successes, failures, support, achievement opened up life's potential. Her loneliness and isolation were no longer adaptive. The feelings they engendered no longer fit. So she put them at arm's length. She disowned them to get on with her life. Unfortunately, by abandoning her feelings she abandoned herself.

That was the origin of the paradox of her defense. By abandoning herself she reconstituted the very situation that led to her feelings of abandonment in the first place. Now the original conflict lived within her. It accompanied her wherever she went. No matter how successful were her new relationships with peers, she could not escape the feelings of being unacceptable. She never felt that she was good enough. As those feelings conditioned her perceptions, the original conflict insinuated itself into every new relationship.

Actually, the little girl she had been, who felt constant feelings of abandonment, was calling out to her developing consciousness. She needed to connect with her feelings by accepting them. She needed to embrace the feelings as if they were a child within, who now could get what she needed only from her.

What child has the wisdom to recognize that? This is why we have parents. They embrace us so that we can learn how to embrace ourselves. If they cannot do that, neither can we. We're left in a world of horrible confusion, where nothing goes right, where our best efforts come to naught.

Finally, it all came to a head. When she was twelve she was already developing an adult body. A boy in her class took to attacking her in the halls of the school. His attempts to molest her were terribly confusing,

---

[91] The symbolism of making dolls' clothes is remarkable, for by doing so she was adorning smaller images of herself. Thus, her unconscious was already trying to transform her situation.

for her father had taught her never to say "no" and her uncle had been molesting her for years. But she was able to avoid his advances.

Then, one day soon after, he hid in the woods through which she walked from school to home. He attacked and raped her. Devastated, she ran home and told her mother. Her mother told her father.

Although both parents assured her that what she had experienced was unacceptable, justice was not to be. The boy was from a prominent political family. His father was her father's customer. Consequently, her father was less than committed to justice for his daughter. The police, not wanting to incur the ire of the local potentate, chalked it up to an unfortunate—but not criminal—event between two teenagers. As no one had witnessed the assault, it was her word against his. Best to forget the whole thing.[92]

She appealed to her mother, who yelled at her to "Get out!" She ran away from home and was taken in by a forty-year-old man, whose motives would be recognizable to any adult. But not to a child. Four days later the police found her wandering on the street and took her home.

She reports that after the rape she was in a fog for several months. She would sneak out of her house at night and go to parties where upperclassmen were hanging out. She became the local sex object. When she arrived at school in the morning, boys would throw pennies at her and call her a whore. Today, forty years later, she is filled with shame when she thinks about it.

It got worse. However predictable from our understanding of the paradox of defenses, no one saw it coming. Once active, no one knew what to do.

In metaphorical terms, the little girl within recognized that however much the experience had harmed her, she had been wanted. Had the boy not risked everything to have her? However erroneous the perception, it dominated her behavior. Her desire had been fulfilled. She was wanted.

She became promiscuous, a party girl who was available to virtually any boy who wanted her. The lonely and rejected little girl within was

---

[92] Unfortunately, this experience is sufficiently common for Jonathon Franzen to include a similar account in his novel *Freedom* (pp. 35ff) that is equally horrifying. To read it is to gain a much more complete sense of the experience than I can provide here.

finally happy. Things were looking brighter. Her problem was solved.

Or was it? Remember, she herself had rejected her little girl's feelings of abandonment. Actually, she was completely unaware that her little girl even existed. That our personalities are comprised of elements other than our conscious egos has little currency in our culture. Often it is considered pathological.

Her little girl was determining her behavior. However gratified her little girl may have been, she herself wasn't. Being the promiscuous party girl gained her little, if any, enduring relationship. She was being used. The paradox was firmly in place: As it usually does, the unconscious sought gratification through the most expedient of means available. That left her feeling consistently and continually abandoned.

Fortunately, fate intervened.[93] She met a schoolmate who was also completely alienated and lonely. Together they would go to parties and look after each other. The rampant sex came to a halt.

However, her feelings endured well into adulthood. She decided to seek analysis. She recognized that her attraction to unavailable (and usually older) men was basically a form of *repetition compulsion.*[94] To her credit, she knew that something was very wrong and sought assistance in figuring out what it was. As we put together a rational, conceptual explanation of the dynamics of the paradoxical defense, she began to understand what was happening to her.

Explanations of what is happening brings a great deal of relief. When patients discover that the highly upsetting circumstances that bring them into analysis are actually quite interesting—even intriguing—and that they present a puzzle to be solved, their relationship with those circumstances changes. Their alienation from themselves mitigates. What was considered wrong and bad, now becomes simply misunderstood.

Jung made a magnificent contribution when he wrote that symptoms are the way the psyche tries to cure itself. The symptom is not the problem. What gives rise to it is. Am I depressed? Perhaps depression

---

[93] Or luck, chance, destiny, God, the devil, the Self—whatever name we give to that which seems to guide us through life.

[94] This term, introduced by Freud, is at the root of the saying that "insanity is doing the same thing over and over while expecting different results."

remedies a toxic situation. Maybe I need to withdraw from the environment. Maybe I need to devote more energy to myself.

These are the kinds of questions that bring the unconscious into consciousness. Understanding does not need to be deep. A woman who was distraught over her psychosis once told me with relief: "I know what's wrong with me! I have schizophrenia." Just having a name for what ailed her reassured her. If it had a name, it was known. If it was known, it could be treated.

Unfortunately, that kind of relief is usually short lived. Having a name for what ails us, even having an explanation for why it exists, does not necessarily transform the illness. Eleanor was greatly relieved to see the pattern in her perceptions and behaviors. As an adult, she knew that they were irrational. Knowing that they were a function of an abandoned little girl within helped her to realize that she wasn't crazy. Once upon a time, they had kept her intact.

Now she could modify her behavior. She could recognize why she was attracted to unavailable men. She could free herself from these serial exercises in futility. She could look for men who were available and who found her attractive. She could stop throwing good money after bad, so to speak. She had hope. She could transform her life.

But it didn't work. She struggled to disconnect from the unavailable men in her life. But every time she saw one, her intense longing recurred. It was one thing to change her behavior. It was quite another to change her feelings.

Trying to change our feelings is rarely successful. They are powerful and insistent. They also have a habit of flooding us when we least expect it, or when we are tired and vulnerable.

Bringing unconscious into conscious is an essential first step, but if it is not followed by bringing conscious into unconscious, success eludes us. We're right back where we started.

That was the situation for Eleanor. Her adult self was being usurped by her child self. Her fears of being crazy were caused by identifying her child self as her adult self. Paradoxically, by attracting her adult self to unavailable men, her child self was forcing her to experience the same abandonment the child self was feeling. The child self was calling to the adult self by connecting her with men she couldn't have.

Back to the father complex. That the child self was replicating the original dilemma (an unavailable father) in the adult self's life forced the adult self to recognize and embrace the child within. The child self needed the adult self to do for the child self that which the child self was seeking from others: To be held, accepted, loved, adored, cherished. But the adult self didn't want to recognize the child self. Moreover, the feelings that the child self generated were disturbing and painful. The adult self pushed the child self as far away as possible.

As we saw earlier, the adult self was replicating the father. The adult self wanted nothing to do with the child self. However, efforts to transcend feelings constitutes suppression. Any psychotherapy that supports the view that our feelings are problematic and assists us in ridding ourselves of them does us a disservice. It increases self-alienation. It makes us feel like failures, for we cannot accomplish the task.

Trying to rid ourselves of feelings prevents the oscillations between conscious and unconscious from bifurcating into new possibilities. It traps us into the oscillating stasis of deterministic chaos. Our mistake is to see feelings as deriving solely from the current situations in which we find ourselves. We don't realize that even if the situations engender such feelings, their intensity is amplified by previous experience. Again, in trauma language, we're experiencing flashbacks.

Eleanor now recognized that the intensity of her feelings had less to do with current situations than with past. Her little girl was calling to her from the isolation of her bedroom, where she found solace only in making doll's clothes. But now it was not her father to whom she was reaching out. It was Eleanor herself. Unfortunately like her father, Eleanor was failing to respond. She could hear her inner child self crying, but she wouldn't go into the room.

Active imagination was the solution to her dilemma. Eleanor had to imagine embracing her crying child whenever she felt like crying. She had to assure her child that she loved her by accepting the feelings that were in the past—and are in the present—her child. However counterintuitive, she had to accomplish the task with no expectations whatever that the feeling would go away. Active imagination in the service of eliminating feelings inevitably fails. Rejection begets only abandonment.

It took her a while, but eventually she got the hang of it. She found a picture of herself as a child and carried it with her. She set aside time to spend with her child. She would conjure up her feelings of abandonment and shame and sit with them as she would with a suffering child.

She began to feel better. She was able to identify ways in which her father was abusing everyone around her, especially through sexual innuendo. Her brother-in-law was always putting the make on her. She decided that the next time would be his last.

She became stronger, more assertive, more protective of her child self. She still felt all of the troublesome feelings, but she no longer felt alienated from them. They were she. By accepting them, she was accepting herself.

Reprise: Synaptic Plasticity

Life experiences become encoded in synaptic configurations. Most often, synaptic dynamics become more complex. Sometimes, entirely new synapses are formed. The brain is plastic. Synaptic configurations that are no longer activated eventually disappear.

Some life experiences are *archetypal* (common to human experience). They occur over and over again. The most basic archetypal experience is relationship. From the moment of conception, morphogenesis involves molecules coming together in new configurations. They make us what we are.

That continues throughout life. Relationships generate the bifurcations and oscillations that foster development. As those bifurcations and oscillations evolve into deterministic chaos, the resulting fractal elaborations bring the profound disruptions that occurred in earlier relationships into new ones. Thus, synaptic configurations that formed in earlier relationships resonate in new ones. We experience the firings of those synaptic configurations not as *then*, but as *now*. Thus we confuse past with present, diachronic with synchronic.[95]

Through active imagination, Eleanor facilitated new relationships

---

[95] For a more complete elaboration of the differences between synchronic and diachronic time, see *Archetypes,* pp. 26ff.

within and among synapses. The old pattern of rejection and abandonment was augmented, supplemented and transformed into new configurations. She literally reconfigured the past. Synaptic configurations created by rejection and abandonment became transformed into new ones, now characterized by her acceptance and relationship.

Through relationship between adult self and child self, relationships within and among synapses formed. Oscillations stuck in the original configuration could now bifurcate into new ones. Stalled individuation could now proceed. The remnant of the past preserved in synaptic configurations could now move into the present.

Metaphorically, Eleanor entered the room where her child self was alone and isolated. She did for herself in the present what she had needed in the past. The child now had all that she originally needed. She could grow up. She could become an adult.

Back in childhood, before Eleanor was conscious enough to understand what was going on, before she was able to analyze it and choose ways of coping with it, the unconscious did it for her. Firstly, it removed her from the situation. She fled to her room. Secondly, it provided her with a symbolic solution to the problem. By making doll's clothes she formed a relationship with a projection of herself. Finally, it made it possible for her to seek in the world outside what she was not receiving at home. Had it not been for the rape, she might have found what she was looking for.

Unfortunately, in extricating her from the original situation with her father, the unconscious also left her abandoned self behind. Those two entities—the self that moved into the world and the self that was left behind—oscillated with each other in the unconscious. Like Sleeping Beauty, the curse would be broken only when Prince Charming arrived.

Prince Charming had to be Eleanor herself. The Eleanor who moved out into the world became her adult self. Her child self remained trapped in the past. Only when Eleanor stopped rejecting her and reached out to her in love was the curse broken. Only then could a new bifurcation liberate the stuck oscillations.

Alice Redux

Remember Alice in Chapter Six? Her scaffold looked like this:

ft: deficient, inadequate
def: does everything for everyone else
res: exhaustion and feeling inadequate, a failure
pp: trying to be overly competent leads to feeling inadequate

Her feelings are generated by oscillations that are stuck between competence and incompetence. Afraid of appearing incompetent, she accepts responsibility for way too much in her life. That sets her up for failure. Simply trying to be competent ensures exactly the opposite.

The result? A tension between arrogance and humility. Her inability to master everything ends up humiliating her. Moreover, trying to master everything is seen as arrogance. Therefore, trying to provide answers to insoluble problems traps her in an arrogance that leads invariably to humiliation. She dooms herself to exactly what she avoids.

Could this dilemma have had its roots in her childhood? She relates that her parents always fought. Home life was permeated by endless conflict. Much of her desire to provide answers to insoluble problems began in her attempts to resolve her parents' conflicts.

That began before she was aware that it was happening. By the time she could think about her home situation and her relationship with her parents, trying to solve everything was an adaptation already in place. Because it was unconscious (unconsciously formulated and unconsciously acted out), she didn't see it functioning in her adult life. All she knew was that she was prey to—as she put it—"compulsive behavior."

The little girl inside her, manifested in her compulsive behavior, was desperately seeking a good parent. Her analysis was one attempt to find one. Another was her quest for credentials. She was enrolled in a degree program that was made all the more challenging by her obsession with performing at the highest possible level. Even more self-defeating was the fact that completion of the degree was unlikely to result in a job. She faced humiliation after humiliation.

Her hope that I would be the longed-for parent led her into spasms of childish moodiness. One moment she would be cute and seductive. The next she would be whiny and complaining. Eventually she would throw a tantrum. The conflicts she had witnessed in her parents began to play out in our relationship. We both felt that I was letting her down.

Or, to put it more accurately, the child within her had those feelings and acted out those behaviors. Her adult self was quite pleasant and related. That I wanted to be the agent of her salvation led me to begin our work together as a good parent who could comfort, advise, support, moderate, challenge, and guide.

Doing so, however, usurped her ability to connect with herself. Helping her was not helping her. As time went on, I would have to encourage her to do what I was doing. She had to connect with her child self. That plunged her into another conflict, namely, how to be both subject and object of her own efforts.

Then the subject of credentials came up. She bemoaned her less-then-perfect performance in her degree program. She lamented its vocational road to nowhere. She began to see credentials as another parent who could not give her what she needed. The school, like her parents, was living off her, giving little in return.

As in childhood, she could not conceive of breaking with the parents. More exactly, her inner child saw no option other than staying with the parents. Her adult side would have to make the break. As she considered giving up her degree program, she sank into gut-wrenching sobs. She felt abandoned, hopeless and inconsolable.

This was the critical moment in her analysis. If she could allow herself to sink into her despair, if she could recognize that this could be a re-entry into that time in her life when she had felt alone and abandoned, if she could allow herself just to be with herself, then maybe she could finally be for herself the parent she had always needed.

It would be as if, when she was a small child despairing in the loneliness of her little bedroom, a loving woman had come in and sat on the bed with her and said: "I love you and will never leave you. I cannot remove you from this horrible situation, but you will never again be alone in it. I shall be with you always." The woman she had needed then was the adult she was now.

As she began to realize what was happening and what was possible now, she cried all the more. But the tone was different. Now, the rage that previously had permeated her crying melted away. She seemed more at one with herself. By allowing herself to cry in hopelessness, she was holding the crying, hopeless child that she had been, had left behind, but was still with her.

The synaptic configurations that had formed in childhood, buried deeply in the unconscious where she could not see them, fired whenever she was in a situation that resembled her home life. As with Eleanor, they were firing not as then but as now. Of course she would identify them as adult reactions to present circumstances. Only when she saw them as reliving her childhood, as being the child she was, did she understand what she had to do. She wasn't crazy. She wasn't disordered. As a well-functioning adult, she could indulge the feelings like a mother holding her crying child on her lap.

Remember the Sidney Harris cartoon in Chapter Three? This was the miracle moment. As she held herself, two things miraculously occurred.

Firstly, her inner child began to grow up. Now with the love and support of a caring adult, she could leave the stasis of her childhood home and enter the world. Isolated synaptic configurations began to interact with the rest of the mind. Every time she fell into her feelings of despair, she fell a little less far. In time, they would be normal adult feelings of frustration and pain. Secondly, her degree program became less of a dead end. The job market had not suddenly improved. Her expectations had changed. No longer did she see credentials as making her a competent adult. Now she realized that credentials simply reflected the competence inherent within her. Credentials no longer legitimized her. Her competence legitimized her credentials.

She began to see that her degree could open doors outside the field she was studying. A myriad of possibilities became obvious once she no longer needed to confine herself to her field of study. She could use her credentials rather than being used by them.

As she tired of her quest for competence, her grades improved. She relaxed and let go. She began to recognize the oscillation between who she had been and who she now was. That recognition allowed the oscillation between adult and child selves to bifurcate into further oscillations—

between competence and incompetence, arrogance and humiliation, hope and despair. Where previously they had oscillated in isolation from one another, now they constituted a chain of bifurcations and oscillations. As deterministic chaos was set free to elaborate, her life began to self-organize.

The Complexity of Miracles

The miracle of analysis is that we never know where it will end up. To open ourselves to ourselves, consciousness must surrender its control. Or at least that's how it feels. In fact, consciousness has very little control to begin with. The illusion of control under whose spell it falls derives from its ability to categorize, define, delimit, rationalize, and conceptualize. All of that simplifies complexity. However complex that may feel to consciousness, it hardly compares with the complexity of the unconscious.

Consequently, much depends on how alienated conscious is from unconscious. The more alienated we are from ourselves, the more turbulent it feels to reconnect. As noted above, most psychotherapeutic approaches seek to fit unconscious into conscious. However useful that may be (for example, by restoring the illusion of control and the security it offers), it is only the first step. The second step acculturates consciousness to the indeterminacy of the unconscious. The unfamiliarity and confusion of that process is best negotiated with the assistance of someone who has experienced it, both in themselves and with others.

Analysis is a journey into the darkness of life itself. Darkness here refers not to some insidious element in the process, but rather to the laborious process of engaging something while resisting the urge to tailor it to consciousness. We must hold conscious and unconscious together while they integrate each other into new configurations. We do so via the images that emerge from the unconscious in dreams, affects and projections. However irrational they may seem, we must allow them to amplify themselves into narratives.

Miracles require faith. In the case of analysis, faith presumes that when consciousness immerses itself in unconscious images, something useful happens. As with religion, we cannot control the outcome. We do not even know how it happens. We may even fail to recognize the change that occurs.

But over time we'll feel better, if only from the increased familiarity of dwelling in the incomprehensible. We'll adapt the limitations of consciousness to the chaos of the unconscious. We'll become more adjusted to the turbulence of life itself. Security will be less a function of control than of relationship. As we become increasingly secure in relationship, relationship will become increasingly our security.

Tom

That is precisely what happened to Tom. Because as a child he had felt that his home was a trap, he left when he entered adolescence. He set up a life of his own that resulted in a measure of success. When he entered analysis, however, his defense of leaving was the only response he could imagine whenever he felt trapped. He always left during those inevitable times when relationship felt like a trap. The paradox was obvious: Feelings of entrapment left him trapped in the defense of leaving.

It wasn't that simple, however. In his family of origin, feeling trapped made him feel like a victim. The desire to leave served to rescue him from that victimization. As is often the case, an internal dynamic (rescuing the victim) became externalized: His security depended on championing the oppressed. That aggravated the already problematic relationship with his parents, for they considered his behavior to be a negative comment upon their affluence—which, in fact, it was. Their defensive criticism left him feeling insecure and unappreciated.

Children need their parents' love and support. No matter how toxic the home situation, children crave their parents' appreciation. Archetypes make use of whatever is available, adequate or not. That can lead to the paradox of the most abused or neglected children often being the most loyal to their parents. Even as adults, they seek their approval.

For Tom, that meant pursuing success. That, in turn, meant leaving home. Thus did his need for his parents' love oscillate with his need to leave home. The latter won out because his defense prevailed over his efforts to connect. As a result, his memories of home were dominated by leaving home as an adolescent. Consequently, whenever as an adult he longed for a home, he found himself moving on. It was as if security necessitated homelessness.

While leaving home liberated him from the pain of unfulfilled relationship (with his parents), as an adult the defense (leaving) consistently deprived him of relationship. The paradox became increasingly complex. By moving on, he essentially never left home.

His enormous inner void eventually brought him into analysis. His presenting paradox reflected the diluting and empowering poles of the defense.[96] Trying to avoid the void only reinforced its existence and ensured its continuing effect. He could not be good to himself, for not only had his parents never shown him how to do that, but he also feared that doing so would compromise his vigilance, whereupon he would slip even more deeply into the void.

With success came a new dilemma. On the one hand, relationship remained a threat, for his defense saw any connection as a liability. On the other hand, success made him attractive, creating new opportunities for relationship. But that simply could not be. Thus, success felt like failure. Openness to others was not an option. The conflict (oscillation) between neediness and the failure of success demanded he disconnect.

Tom turned to self-sufficiency as a possible solution to the conflict. Maybe self-sufficiency could give him a secure footing outside of the quagmire of relationship. Unfortunately, his attempts to be self-sufficient deprived him of others' collaboration and assistance, the result of which increased his feelings of failure and loneliness.

Fearing abandonment by others if he was "too big for his britches" (that is, thought too much of himself), he rejected anything about himself that was positive or ego-enhancing. Fear of rejection by others led only to self-rejection, which left him feeling abandoned. As far as the defense was concerned, however, that had been his most secure identity from the beginning.

Leaving infected everything. He abandoned his parents, his home, success, others and himself. Previously when he had sought therapy for his feelings of abandonment, he had been encouraged to adopt ways of

[96] See Chapter Four, The Chaos of Defenses.

making them go away: His cognition was restructured; he was stimulated bilaterally; he was made more mindful of events and his reactions thereto; he was encouraged to activate his higher powers; he learned to desensitize himself systematically— all to no avail.

Therapy of that kind didn't work because it focused on abandoning his feelings of abandonment, on rejecting his feelings of rejection. The paradox of his defense was amplified by the paradox of psychotherapy. He was trapped in a loop. It was as if he had never left home.

Projection and transference permeated his various therapies.[97] For example, just as entrapment and rejection characterized his relationship with his parents, psychotherapy trapped him into rejecting his feelings of rejection. It felt right, but for all the wrong reasons. Psychotherapy felt like home, albeit with a novel element of the caring and concern of the practitioners who were doing all they could to help him out.

Out, however, never occurred. There was no metaphor, no image for what was occurring in the defense. Once he entered analysis, however, the metaphors and images spontaneously appeared. When in the consulting room Tom experienced the strong and overwhelming feelings of entrapment and rejection, he was able to see that they were not a function of his relationship with his analyst, for they were completely disproportionate to whatever entrapment and rejection may actually have existed in the analytic container.[98]

Obviously they were connected to something else. He recognized right away that they derived from childhood. What he did not realize was that they were not simply memories and recollections. They were also flashbacks. Those flashbacks were not simply digging up something from the past. Rather, they were the reconstitution in the present of a present that had occurred in the past.

---

[97] Projection refers to seeing in others unconscious aspects of ourselves (often our own shadow). Transference refers to interacting with others in ways consistent with how we were taught by our parents.

[98] Analysis, for all of its differences from everyday life (boundaries, for example), is essentially a microcosm of everyday life. Consequently, submitting to the wisdom and experience of the analyst can feel like entrapment. Likewise, the analyst's failure of empathy or resonance can feel like abandonment or rejection.

In other words, his feelings were those of his child self that continued to exist in a home his adult self presumed to have left behind. They were the lonely, abandoned and trapped self that came into being in childhood and continued to exist in adulthood. He may have left home, but his child self was still there. Again, past was present and present was past.

In analysis he could change the past by intervening in the present. He could leave diachronic time and enter synchronic time. What he had needed as a child and what his child self needed now, he could now be. Like Eleanor, he needed to embrace his feelings, not seek their disappearance.

We practiced in the consulting room. My role was to model to him how to embrace his feelings. Unlike his parents, I needed to accept and value his feelings. The analysis required me to be the parent he never had.

This is not always easy. I have never particularly enjoyed being the focus of rage, disappointment, accusations, and insults. Furthermore, child selves can be immature and manipulative. For good reason do we suppress them. There's not much room to maneuver in the narrow space between genuine and manipulative vulnerability.

I had to be sympathetic with some needs, but not with others. I had to admit responsibility for some of the difficulties in the analysis, but not for others. I had to encourage relationship in some ways, but not in others. Perhaps most importantly, I had to tolerate the ambiguity of the analytic process and my ambivalence in participating.

Fortunately, the analytic container is well designed to accommodate the complexity of the analytic process. Analysts are able to enter into deeply intimate relationships with patients because certain aspects of intimacy are proscribed. The analytic relationship is limited to the analytic container. We do not socialize with patients outside the container. We do not form romantic relationships with patients either inside or outside analysis. We do not enter into business relationships with patients (except, of course, with regard to reimbursement for our services). We share little—if anything—about our private lives.

All this makes intimacy in analysis necessarily abbreviated. While patients are right to expect analysts to facilitate their individuation, analysts must never be the agents of it. That role belongs to the relationship between the patient's conscious and unconscious. Analysts are more like midwives. We facilitate something that has been put into motion by something other than ourselves.

Consequently, however I might be able to connect with Tom's child self, eventually Tom had to assume the task himself. I could only go so far. I could show him how to do it, but I could not do it for him. We could practice in the consulting room. Outside, he would have to do it himself. Analysts' inadequacies can sometimes be very useful.

And he did. He understood that he would have to forego the kind of conscious appropriation of unconscious that had previously characterized his therapy. Instead, he would have to submit consciousness to the uncertainty and imprecision of the unconscious. He would have to enter the chaos.

He became acutely aware of when the intensity of his feelings indicated the presence of his child self. Rather than either suppressing the feelings or berating himself for having them, he would imagine himself holding them, listening to his child self, comforting and accepting him.

He discovered that holding and comforting his child self led him to feel comforted as well. What he had lacked at home he now had. He was on the receiving end of his offering. Adult self and child self could become one through relationship with each other. Who he had been at home was now who he was. Their shared wound could heal.

From a mechanistic perspective, introducing a new bifurcation into the oscillating system liberated it from impotent oscillations. Previously, Tom's psyche had become stuck in a system of oscillating paradoxes (for example, trapped or leaving, success or failure, relationship or self-sufficiency) that could not progress beyond them. The alienation of adult self from child self created the psychological analog to a "Mexican standoff" that entrapped them in a dynamic that was going nowhere.[99] Essentially, they were still trapped in the home.

[99] The expression "Mexican standoff" describes two armed individuals who are facing each other. Each one strains to notice any movement on the part of the other. As a result, whoever first makes a move, any move whatever, will alert the other to shoot. Thus, both will die if either makes a move. So, they simply stand there, facing each other, paralyzed.

Tom's child self was evidence of that situation. Analysis gave him a metaphor—an image—that made it possible for him to redeem it from its paralysis. That was the first movement in the analysis: Tom became aware that his feelings were himself at an earlier age. Conscious deciphered unconscious.

The second movement required Tom to submit himself to the unconscious dynamic that was giving him trouble. He had to relate to it as a valid dynamic that could provide a solution to his dilemma (the oscillating system of paradoxes, mentioned above), however unpredictable and uncontrollable that might seem. He had to surrender himself to non-rationality. Conscious had to surrender itself to the non-rationality of unconscious.

Summary

Destruction and healing are both a function of relationship. When relationships in early childhood are so destructive that they threaten the health of children, the unconscious forms defenses to improve the child's survival. Those defenses become locked into the child's perceptions and behaviors. They form feedback loops that reinforce the defense.

Deterministic chaos describes the process well. There is sensitive dependence on initial conditions, iteration and mutual inhibition. Where the damage begins has a great effect on where it ends up. The defense becomes synchronic and acts like a constant in life's equation. Perception and experience feed back into each other, maintaining the original situation for which the defense was created.

The defense dynamic is fractal: Self-similar across scale and transcending categories (unconscious). Consciousness exists in diachronic time—past, present and future. The unconscious defense inhabits synchronic time—ever-repeating, ever-present, ever-the same. Then is now and now is then. To treat the defense as limited to events in the past is to deprive ourselves of the opportunity to revisit the past in the present. We cannot transform ourselves if we are lost in either past or present.

Deterministic chaos bifurcates and oscillates its way into eternity, producing the same patterns over and over. Like images from the unconscious, we can vaguely perceive the pattern but never reduce it to its particularities. We glimpse them only through a glass darkly.

But if we treat it all as a present reality in synchronic time, we can enter into the dynamic that gave rise to the defense. We can relive childhood. We can become for ourselves that which, had it been available in childhood, would have obviated the need for the defense. Once conscious has extracted sufficient information from unconscious to understand the challenge, it can throw itself back into unconscious to participate in the original drama that activated the defense.

That is the secret of transformation. It requires a cooperative relationship between conscious and unconscious. Through that alliance new bifurcations can be brought into relationship with the chaos of the defense.

We'll devote our next two chapters to ways through which to accomplish the task. In the next chapter we'll look at autoimmune disorders. Finally, we'll look at the myriad ways in which active imagination can be engaged. With that, we'll have completed our description of how to transform the strange attractors of childhood trauma

8
Autoimmune Disorders

Into the destructive element immerse![100]

What's it like for conscious to immerse in unconscious? The following two examples focus on autoimmune disorders. Here is an excellent object of our investigation, for the health sciences understand very little about them. Current explanations of the causes of autoimmune disorders range from the immune system mistaking our own molecules for those of pathogens, to the immune system arriving at the site of tissue damage from toxic chemicals and, finding no pathogens to dismantle, dismantles our own tissues, to reduction in threats (elimination of parasites, for example) leaving a hyperactive immune system overreacting to minor insults.

All are hypotheses yet to be proven. None explains what actually happens. So the situation is ripe for speculation, which is exactly what we shall do in the following examples. Please do not mistake anything you read herein for good medical advice. While medicine may know very little about autoimmune disorders, irresponsible speculation hardly provides a viable alternative.

By definition, explanations devoid of precision have failed to be differentiated from unconsciousness. The following examples try to envision new bifurcations that can unlock oscillating systems that appear to be going nowhere. The goal is to create bifurcations with deterministic chaos that can generate new fractal patterns. In other words, we are trying to activate morphogenesis.

We'll begin by assuming that the best way to approach something that resides in the unconscious is via the unconscious. With regard to autoimmune disorders, might it be possible both to engage and to transform them through bringing consciousness into unconsciousness?

That's the opposite of how it's usually done. But it is the core of Jungian analysis. We've certainly seen how it works with the strange

[100] Stein's words to Marlow in Joseph Conrad's *Lord Jim,* p. 214.

attractors of childhood trauma. Now let's see how it relates to autoimmune disorders, whether or not they share the same origin.

For example, consider what transpired when the following individuals entered analysis.

Ruby

Remember Ruby from Chapter Six? When she first entered analysis her life expectancy was about six months. She was suffering from an autoimmune disorder and was not responding to treatment. The problem was compounded by her medication, prednisone, which is a powerful steroidal hormone. Too little, and her disorder compounded. Too much, and it would eventually kill her. She was truly caught in a conflict between illness and treatment.

We explored the possibility that her autoimmune disorder was a fractal elaboration of a strange attractor that formed as a result of childhood trauma. When she was six, her father had died. He had been the more emotionally available of her parents. While her mother loved Ruby and was a good person, she was also preoccupied with her career. When Dad died, the mothering element in Ruby's life essentially disappeared. Ruby was devastated.

Further complicating the family dynamic was the need to earn money. Father had been the breadwinner. Now mother had to fill that role.

She was very good at it. She became nationally known in her career and supported the family very adequately. But her career required her to spend extended periods away from her children. Taking care of the home was left to Ruby and her older siblings.

They did well. They developed camaraderie, cooperation and resilience. They became self-sufficient and industrious. They did well in school. They triumphed over adversity.

We weren't sure how that related to Ruby's disorder, but it seemed likely that there was some connection. The death of a parent is a severely traumatic experience regardless of the eventual outcome. That we may function well afterward does not mean that we are not traumatized.

We tried a number of interventions. Like her mother, Ruby was a professional. She was well liked by her friends and colleagues. But she had never married. As relations with men were infrequent and unsuccessful, we looked at Ruby's impressions about men and dating. We explored ways in which she could adopt new perspectives and find new opportunities. Some of these paid off for the short term. None, however, took.

Was that a result of childhood trauma? She had a recurring nightmare that seemed to suggest so. In it she was pursued down a long, dark hallway by a man who was intent on raping and murdering her. She would awaken just as he reached out to grab her.

We tried dreaming the dream onward. She would imagine what would happen in the dream if she didn't wake up. As we are never injured nor do we die in a dream, she could keep the narrative going. She imagined herself out-running him, or escaping his grasp, or having the whole thing evaporate in her mind.

That helped. The nightmares came less frequently. But it didn't eliminate them. We needed something else.

So we tried hypnosis. Ruby would put herself into a light trance through a relaxation exercise.[101] As she relaxed I would count backward from five and ring a bell at one. While that may seem rather Pavlovian, it did help her to give up conscious control and to surrender herself to her unconscious. She imagined that I was there with her and that we both could confront her attacker.

What did he want? Why was he focusing on her? Was there something she was avoiding that she needed to engage? Who was he? What did he need from us?

We received no definitive answers. But how could we? That would be the domain of consciousness. We were immersing ourselves in the unconscious. There, we received not answers and explanations, but feelings and impressions.

Ruby developed an interest in her nightmares. What had previously been a source of terror now moved in the direction of intrigue. She could be a participant-observer in her nightmares rather than trying to escape. She felt more in control even though she was actually surrendering it. Another paradox—in this case, quite useful.

---

[101] Please see the appendix here for an example.

Then came the breakthrough. One night as the nightmare recurred she suddenly stopped running and turned around to confront her attacker. It felt perfectly normal. She was not terrified, just interested. She stopped and turned around.

There before her stood a little boy. She estimated his age at about six years old. He was lost, frightened and crying.

Her heart went out to him. When she asked him what was wrong, he replied that he was lost and couldn't find his way home. Then she awoke.

In our next session she could hardly contain her excitement. We both immediately celebrated her discovery. Now the dream made sense. She had lost her father when she was six; home life had been compromised, she had lost her way and had never been able to find it again.

Then it occurred to us that the dream was not simply recalling metaphorically what had happened in childhood. The little boy was—and continued to be—her as a child. The synaptic configuration that formed in childhood was still intact. Whenever it fired, it fired as *now*, not as *then*. That little boy was part of her right now, in the present. And he was still lost. Whatever she accomplished on her way into adulthood had not included him. He had been left behind. Once abandoned by the death of her father, now he was abandoned by her.

Explanation led to interaction. Progress would now be made not in the land of cognition but of imagination. The dynamic existed in the unconscious. It would have to be transformed there. Dragging it into consciousness and subjecting it to rational dissection would cripple the boy.

So Ruby concentrated on embracing the little boy. To do so, she surrendered herself to his realm—the unconscious. She told him she would help him find his way home. In fact, they both found a home. Together they found their way back to something that had existed before her father died and to some extent had never ended. She developed a deep love for her little boy. She imagined him being with her during the day, at home and at work. She became happier, more secure, more related, less dependent upon control.

When she enfolded him in her arms in bed at night, a very interesting thing happened. The nightmares ended. What had pursued her had found her. She had accepted and embraced herself. That greatly relieved both of us.

But there was more. The next check of her prednisone levels revealed that she was considerably overmedicated. The dosage was halved. Over time it was halved again. The "six months of remaining life" was revised. Now, there was no definitive estimate.

The moment Ruby began to engage her unconscious through active imagination, her inner psychological dynamics changed. She created a new bifurcation with a deterministically chaotic attractor left over from childhood. Her psyche was liberated from stasis to progress. Through the two-way street of unconscious into conscious and conscious into unconscious, she got her life back.

While active imagination often involves the body, with autoimmune disorders it becomes a prime focus. Ruby associated her inflammation with heightened vigilance against attack, from both her body and her nightmares. Consequently, working with her nightmares constituted working with her body. On the psyche-soma interface, dreams can sometimes be spontaneous eruptions of psychic images that accompany somatic processes. Such dreams are essentially imaginal traces of bodily processes.

Ruby's autoimmune disorder did not go away. This was not a miracle cure. Rather, she changed her attitude toward the disease. The feared attacker who pursued her was actually a lost little boy who was searching for his parents. Once she assumed the parental role, the disorder settled down. It was no longer lethal. It was simply one more brick on the load of life.

Chuck

The primary goal of analysis is to bring the various dimensions of the psyche into relationship with one another. Whatever positive results occur are hard to predict. To enter analysis with preconceived notions of what should happen is to grant priority only to that which can be consciously formulated. That heightens one dimension, that of the ego and consciousness, over others (the unconscious). Right from the start that establishes an imbalance in the psyche, which is the very condition that brings patients into analysis to begin with. Such analyses have a hard time getting going, for the ego's grip on the process must first be loosened.

Chuck already knew this. He had been interested in Jung for many years, had read much of Jung's *Collected Works*, and regularly attended Jung Society meetings. Thus, when he entered analysis he was fully prepared to open himself to his unconscious.

Chuck's presenting problem was an autoimmune disorder that has no known cure. In most cases it is fatal.[102] When it had been diagnosed thirty years earlier, treatment sent it into remission for twenty years. For the last ten years, however, he had been trying just to stay alive. He had been hospitalized often and had almost died a number of times. He had been in psychotherapy twice before but had accomplished "nothing."

Currently he felt like "there is a rat chewing at the base of my brain." He felt that the immune system was eating away at healthy tissue. His throat was always dry, his sinuses consistently leaked profuse amounts of pus, he suffered small strokes, his lungs were compromised, and his joints were swelling. Kidney functions were down. He was ravaged by vasculitis—inflammatory destruction of blood vessels While he had been given many different treatments, none seemed to be working.

Nevertheless, he was receiving very good care. His primary care provider was an "old-school" physician who encouraged Chuck to do as much research as he could on the disease, took the time to discuss what Chuck had learned, and respected and valued his input. That practitioner was as essential to Chuck's survival as were any of the tests and medications upon which modern medicine relies.

We spent our first sessions getting to know each other and becoming familiar with the landscape of Chuck's psyche. We accomplished the latter through three sessions of free association, during which Chuck allowed his mind to roam freely in thought and feeling. His job was simply to report. Editing, masterminding, or censoring were set aside.

Free association is often seen as an exclusively Freudian technique. Nevertheless, it has an important place in Jungian analysis. Firstly, it

[102] Untreated, the mortality rate at one year ranges from 82-94 per cent. Treated, at five years the rate drops to between 77 and 79 per cent.

helps patients to get a feel for allowing the unconscious to speak unedited by the tools of consciousness (description, definition, categorization, conceptualization), each of which leaves out more than it includes.

Secondly, analysts are very intuitive and love to engage what patients bring to the session. But as we know next to nothing about patients in the first few sessions, we should withhold our observations until we know more about what they are saying.

Finally, everyone's mind self-organizes. Through free association, patient and analyst see how the patient's mind is organized. What is connected to what? What that may appear to be connected actually isn't? How are emotions interwoven with images? Free association allows us to wander together through the landscape of the patient's mind.

We discovered that Chuck's childhood relationship with his father had been positive. With his mother, it was not. His older brother and he had frequent childhood conflicts. They fought a lot. When Chuck was seven he left home and took up residence in a local swamp. Currently his relationship with his mother is good. His father and brother are deceased.

Dreams were an important part of Chuck's everyday routine. He was skilled at remembering and reporting them. They were a living, breathing part of his life. As a result, our work began to focus almost exclusively on dream interpretation. We explored how elements of the dreams corresponded to his waking life and employed active imagination to bring waking life back into the dreams.

Welcoming the unconscious through valuing his dreams generated a collaborative relationship. We became a team as we walked together in the penumbra of the unconscious. Together we shared amazement at what was revealed. We felt very fortunate to be able to receive and experience the inner workings of the psyche.

That in itself was healing. No longer was the disorder solely a threat and a burden. Now it was an object of fascination, a mystery to be explored. More importantly, we had an ally in our search for answers: the dreams themselves. It was as if the unconscious was teaching us all we needed to know. We were the students who were excited to be able to sit at the feet of the master.

That is analysis at its best. It's why it can become the most engaging activity in life. Nevertheless, we subjugated to the unconscious neither our will, nor our discernment, nor our option to decide how to integrate its offerings into our daily lives and work. We adapted conscious to a dialogue with unconscious. Neither held sway. The collaborative relationships that we developed with each other reflected the collaborative relationship between conscious and unconscious.

In our first session he presented a dream of a greenhouse full of strawberries. Under the greenhouse lived an oriental woman. She was somehow connected with the ability of the strawberries to grow. As we were at the end of the session, I asked him to actively imagine the woman to see what would transpire.

In the second session Chuck reported that in his active imagination she resembled a woman in another city who has the same autoimmune disorder. He felt close to both women and felt that their presence in his life was a blessing. When in active imagination he had approached the woman under the greenhouse, she did not withdraw. He asked her what she was doing. She replied that she was tending the strawberries.

Our view of the dream was positive. We felt that it indicated a nurturing element in the unconscious that could promote growth. I encouraged him to see his body less as an enemy than as a friend, for it had kept him alive throughout the challenges of his autoimmune disorder. I asked him to thank his body three times a day. At first that seemed inconceivable, for he felt his body was destroying him.

His associations with strawberries were that he likes them, they are flavorful and nutritious, and somewhat delicate. At that point we decided not to analyze the dream any further. We preferred to wait to see what else might appear. How would the unconscious continue the dialogue?

Several sessions later, he brought in a dream that had occurred a month before our first session:

Dirt and Water

A man is desperately searching the ground. He is in danger of dying. He is looking for a particular kind of dirt. He finds it and throws himself down onto it and starts to cover himself. It is light-colored soil and I am mildly surprised because I thought dark-colored soil was more fertile.

As he is covering himself (he is face down) water from a stream starts to pour over him from the hillside.

A competent-seeming woman rushes to help him cover himself with the dirt. As water is cascading over the dirt, over his head, she quickly takes what looks like a guitar pick out and scoops a sample of the soil. I am puzzled as to why she is doing that.

Then I see her run along the bank with the sample in front of her looking for soil that matches. I then understand.

The man knew exactly what type of soil he needed to save himself and she knew to trust him and look for what he needed.

Running along the bank, she finds a place where some work had been done before and brushes away the dark dirt to expose a large amount of the light dirt. She quickly scoops up a large quantity of the light dirt and rushes to the man and helps to cover him even as water and dirt are already covering him.

I wonder how he can live covered with water and dirt but I realize there must be some wisdom in this that I am not aware of.

Now the dreams were really taking hold. He connected the dry dirt to his mother, which suggested that his Mother Earth landscape lacked the spiritual principle of water. In the dream a woman is central to the narrative. As in the previous dream, the anima was on his side.

There are also a number of contradictions to Chuck's assumptions. Seeking the light-colored soil flies in the face of Chuck's experience of dark earth as the more fertile. The dying man in the dream could possibly symbolize his own fear of dying from the autoimmune disorder. But how can burying oneself alive possibly be anything other than assuring one's own death? Isn't being buried under earth and water a sure sign of death? What about the guitar pick with which the woman takes a sample of the dirt? Why is she doing that?

In the dream Chuck realizes that the man knows exactly what he needs and the woman trusts his judgment. That comprised two very optimistic elements. Firstly, Chuck is aware in the dream of what is happening. He is not solely a detached observer who simply remembers and records them. He is present in the dream. He sees what is happening and adjusts his assumptions to fit his perceptions while he is dreaming.

Secondly, he sees that what appears to be counterproductive is actually the solution to the problem. That suggested that his orientation to his autoimmune disorder was off. Where previously he had second-guessed everything that his practitioners advised, now he realized that his own understanding may not be sufficient to guide his treatment. He soon relinquished his obsession with his treatment and left it to the practitioners. That, in turn, freed him to consider how to improve his quality of life.

Both established bifurcations that could lead to a more complex oscillatory system. Chuck was trading his obsession with death for possibilities for life. He was being reborn.

Shortly thereafter, two dreams portrayed the profound shift that was occurring within him:

> *The Battle of Brothers*
> I am in a wilderness. I am fighting with my brother. He is older and more powerful than I am but I skillfully outmaneuver and control the fight and ask him to concede. He will only submit under brutality and excessive force which I refuse to invoke. And the fight continues to each moment of my mastery over him and his refusal to concede. (The dream seems to resolve but I forget how. I think he finally concedes. It is some ending that felt complete and good.)

> *A Fight Where There is None*
> I am fighting with my brother's wife on the same Alpine slopes where he and I fought. She indicates she doesn't want to fight but I find it hard to accept. After a short time I realize I have no fight with her and I stop.
> She is in this place to enjoy a vacation and life. I being very familiar with the area start telling her about things she can do. (I think about my real-world relationship with her and think how unfair it would be to continue my struggles with my brother with her now that he is gone.)

Conflict characterizes these dreams. That they involve Chuck's brother and sister-in-law indicates that they are intimately associated with the dreamer. In childhood he had frequently fought with his brother, usually at his brother's initiative. As his brother was older, Chuck usually lost.

That reminded us of his autoimmune disorder. There, he was barely holding his own, if not on the losing end of the battle. In the dream, however, he has the upper hand. He could be the victor were he capable of brutality. Instead, he chooses to control the fight until his brother concedes. Eventually, the conflict resolves.

The conflict with his brother's wife ends when he accepts that she doesn't want to fight. At that point, their conflict disappears. The bifurcations mentioned above now take root in the shift from conflict to cooperation. The desire to enjoy life and take a vacation supplants the conflict. Chuck then observes that the resolution in the dream should extend to waking life. Inner and outer conflicts are being replaced by enjoyment of life.

Chuck was not yet ready to thank his body for surviving the disorder. But he was seeing its symptoms less as enemies than as something he had to endure. He didn't have to capitulate to his body. He could move beyond it, treat it as a necessary burden, yet not allow it to interfere with his desire for a higher quality of life.

The sixteenth session was a breakthrough. Previously we had used his dreams to inform his relationship with his disorder. Now, however, we were beginning to see the dreams as commenting directly on our work together. It was as if the unconscious was showing us the way to healing and strength. At the beginning of the session, he presented the following dream:

> *The Fisherman's Friend*
> I am a person who swims out into a river to place the baited hook at the end of the fisherman's line in the best place to catch fish.
> As I am briefly waiting for the right moment I become hooked myself and the fisherman tugs as though he has a fish.
> I yell or communicate that he has hooked *me* instead of a fish and somewhat to my surprise he stops tugging. (I am surprised that he heard me.)
> It is *very interesting* to me to discover that the fisherman is listening to me.
> I swim out and place the baited hook in the stream in the hopes that he will catch a fine fish.
> I don't see a fish but somehow I know this is how it works.
> (I awoke from the dream feeling very good about the dream.)

We decided not to do an extensive analysis of the dream, but to focus instead on how it affected Chuck. In this case, he felt that the fisherman was God. Where previously he had been "hooked" by God in frightening and painful ways, he felt that the dream was revealing a transformed relationship. Now he was "off the hook," so to speak. He and God were cooperating in a manner that could benefit them both.

Jung believed that the best images for representing the Self were images of God.[103] While Chuck's interpretation is preferred, it is important to note that the dream indicates that his ego's relationship with the Self is also improving. The collaborative relationship between conscious and unconscious is strengthening.

Could this be significant with regard to his autoimmune disorder? While it is very precarious to misconstrue correlations as causations where psyche and soma are concerned, we do well to keep in mind the dynamics of the psychoid realm of the psyche. As we noted in Chapter One, the psychoid realm of the psyche is where body and mind meet. As it was, this dream inaugurated a series of dreams that were to have a profound effect on Chuck's relationship to the disorder. The sought-after fish was appearing. At the beginning of the next session, he presented the following dream:

The Healing of Elk-Woman

Outside the window of my parents' house a presence is felt. I notice some movement. I see that it is an elk or caribou. I get my bow. My mother motions for me to hesitate. I do. I then notice there is something different, something vulnerable, about the movements. I put the bow away. I watch. We watch.

I notice it is a woman and an elk wearing a blue dress. In the fashion of an Indian woman but with this incredible slightly metallic indigo blue color. Something appears to be wrong. Upon noticing us such a wild creature should have moved away from the house, but she is staying in this back corner for some reason.

I go out and I see that she is pregnant and in distress. I move quietly closer to comfort her. I worked with animals a lot as a kid and was always able to bring comfort to them. They knew that they could trust me and I want her to know she can trust me. She does.

She lays down along the outside of the house and a bit to my dismay I realize that I am to help her give birth. I think someone else should do it

---

[103] "The self is not only the center but also the whole circumference which embraces both conscious and unconscious; it is the center of this totality, just as the ego is the center of consciousness." *(Psychology and Alchemy,* CW 12, par. 44); "The self is our life's goal, for it is the completest expression of that fateful combination we call individuality." *(Two Essays on Analytical Psychology,* CW 7, par. 404)

who knows how but there isn't anyone else—just like on the farm when I was growing up.

She gives birth but instead of a baby it is a misshapen growth. What looks like brain and spinal cord and other tissue. I clear these things away and comfort her in the process. The most important thing is to help her. I clear away most of the tissue and see some of it is still attached to her and I have this idea to cut it away to further complete the process and clean up the situation but a deep voice, a knowing, a presence tells me that it is okay to leave this part. That to take more would be harmful to her. Just leave this and, as in days of the past so in the near future, these things will be absorbed by natural processes. Continue to comfort her now and things will be all right.

I follow the advice of the voice and turn my attention to making her feel comfortable and safe. I worry that she will become afraid but she does not. I am finished cleaning her up and pull a human dress over her body. It seems strange because she is an elk but she is just fine with it and goes into a state of deep relaxation. She knows I will take care of her while she rests.

Soon she is up and hanging around while she recovers her strength.

(The dream is very moving and takes me back to my childhood and all that was good about growing up on a farm. I'm flooded with memories about life and filled with gratitude for this part. The part about the unexpected things and the inherent wisdom built into living things.)

In this dream we saw a clear shift in Chuck's attitude. In the beginning of the dream he intends to kill the invading presence. But his mother intervenes and advises him to hesitate. As they both watch, the image evolves from an animal (elk or caribou) to a theriomorphic image of a woman-elk. Chuck especially notes her vulnerability.

We did not do an extensive analysis of the dream, but focused on its effect on Chuck. He was struck by his shift from hunter to helper, as well as by the external voices that advised him (his mother and "a deep voice, a knowing, a presence"). We also observed that his role in the dream reminded him of his childhood, before he was stricken with his autoimmune disorder.

The birth is also not what Chuck expected. Rather than a baby, a *misshapen growth* is born. Very important is Chuck's assumption that it needs to be cleaned up, cut away, eliminated—an assumption that is countered by a voice telling him not to do that, but simply to care for the woman-elk. He is not to try to remove the apparently problematic element, but simply to let it be. Other forces will take care of that. He need only provide comfort and solace to the woman-elk.

This dream was to mark an important shift in Chuck's attitude toward his disorder. Rather than trying to eliminate it, he realized he needed to take care of his body. The idea of gratitude toward his body that he found so counterintuitive at the beginning of the analysis was now taking hold.

During this time he discovered in his reading of Marie-Louise von Franz that making peace with an animal constitutes making peace with instinct. His active imaginations with the dream helped him to adopt a much more relaxed attitude toward his health. Opposition and alienation were being replaced by comfort and safety. He was learning to trust the "unexpected things and the inherent wisdom built into living things."

He was especially interested in learning about the psychological dimensions of his disorder. During his active imaginations with the elk-woman, his medication was changed to a newer, gentler one. He was also becoming very engaged with the analysis.

Were active imaginations stimulating morphogenesis? If so, was there any evidence appearing in the unconscious? A month later he had the following dream:

> *Saving the Eggs*
> The eggs have been scattered and she needs help. They [the eggs] are aliens but somehow connected to me, connected WITH me and I feel moved to help her. It is my responsibility and desire.
>
> Other people don't care. Someone even tries to boil some of the eggs and I take them back and give them to her. The eggs need water. They need water or they will not survive. I put them into a rectangular case [the eggs are rectangular] that is an old computer case and put a water hose on them spraying them with water.
>
> Soon I tell her that this isn't going to work for very long, that we have to get them back to sea water. The hose isn't providing a sufficient quantity of water and sea water is better for them.
>
> (I'm actually left feeling somewhat optimistic from this dream. The eggs are important and it feels like the right things are being done for them now. Although it has been somewhat traumatic to have them scattered and

boiled and almost lost, I feel confident that they are strong enough, healthy enough, to survive and that I will get them back to the sea water. Interesting that the dream does not feel complete. The eggs are NOT in the seawater YET.)

Once again, Chuck's read on the dream was positive. As with the elk-woman dream, he could have focused on the negative—*then*: what wasn't born properly; *now*: the inadequate water. But he was intuitively optimistic. The deep voice in its knowing presence was becoming a trusted ally. Facts were being transformed into potentials. How things were was becoming how things seemed.

Chuck's orientation to having an autoimmune disorder was changing. It was being transformed into an active involvement with his body in the face of its challenges. Chuck and his body were facing them together.

While Jung was clear that images rarely have a consistent and uniform symbolic meaning, certain images do stand out. Water, for example, often—if not usually—symbolizes the unconscious. Eggs are similar. They tend to suggest the potential for new life. The shell of the egg protects that potential.

That the eggs are rectangular is interesting. The strength of an egg shell derives from its ovoid shape (*ovoid*, from Latin *ovum,* egg.) Rectangular eggs presumably would not be as sturdy. However, they stack well. This makes it easier for Chuck to protect them, in an "old computer case!" That, plus the fact that the eggs are "aliens," led Chuck to associate them with a major project of his: To create a website about his disorder for people in China.

She and *her* are not named. Chuck connected with her in the dream but had no associations from waking life. She was clearly associated with the eggs' vulnerability and needed his assistance, which he was happy to provide. *Other people* are a problem. But he surmounts it. They simply do not understand the importance of the eggs.

Chuck knows that the water he is using to preserve them is only a temporary fix. They need sea water. Symbolically, the sea is the Great Mother of us all, for all life came from the sea. Chuck was painfully aware that fish stocks are being decimated into extinction by rapacious profitmongers and desperate fishermen. He knew that as go the oceans, so go we.

Thus, this dream has a dual focus. As with virtually all dreams, it pertains to Chuck's personal life and to that of the collective to which he belongs: The human race. The dream is a message in the form of a snapshot.

So here we have it. Chuck's dream reflects his desire to make a contribution to the human race concerning his experiences with his autoimmune disorder. All of his intensive study of the disorder serves not only his own healing, but also that of people throughout the world, whether they have his particular disorder or some other. And he has found an ally. The elk-woman now morphs into a generic female who also needs his help. Chuck has been called. He has been given a vocation.

Or, to put it more exactly, his self-chosen vocation has been confirmed by the unconscious. He is to become an expert on his autoimmune disorder, not in the sense of deciphering its etiology, its chemistry and bodily ramifications, but its significance for those who suffer from it. Chuck is to be a healer who shares his knowledge with others.

The dream "is not complete." He has more work to do. What might that be? A month and a half later he had the following dream:

> *Clearing Out the River Bed*
> I am working on a huge project of clearing out a river bed getting ready for a stream of water. There are trees to clear, debris. There is even some kind of rotting turkey that I find, and my brother gets on a backhoe to dig a hole for it and just bury it.
> (I wake up feeling energetic and get busy with the things I want to do today. It's a LOT.)

Chuck researched possible meanings for the images in his dream. For him, a "psychological meaning" might be that "a river may represent the flow of the life force. In a spiritual sense, it may show your acceptance of divine will and destiny. Instead of struggling against life you 'go with the flow.' Crossing a river may symbolize a fundamental change of lifestyle."

For a "mystical meaning" he found: "Consider your dream in the light of this quotation from *Siddhartha* by Hermann Hesse:

But he learned more from the river than Vasudeva could teach him. He learned from it continually. Above all, he learned from it how to listen with a still heart, with a waiting, open soul, without passion, without desire, without judgment, without opinions."[104]

Chuck decided to clean up the debris in his life. Literally. His home and land were crowded with debris that he had neglected while he poured his energy into researching his autoimmune disorder.

Now that was to change. He threw himself into his work and put his concerns about his disorder on hold. Each week he reported the progress he had made and how it was energizing his life. He was waking up in the morning looking forward to the day, where previously he had dreaded it.

That he had made the correct decision was reinforced by two very powerful dreams that he interpreted to apply directly to his disorder and how he was coping with it:

Somehow a can of gas has accidentally been torn on top and lit on fire. Everyone is scared and I grab the can and put my hand over the spout and over the tear.

The fire is still going inside the can and we are still concerned it will explode. I need to find a place to throw the can away.

A man goes with me as I run to a place that was a field or a dump but now has kids playing basketball there.

I can't throw it there because it would probably burst into flames and spread fire.

I don't know where else to go so I stand there clamping down hard on the holes in the can, hoping to smother the fire before it explodes.

(My first thoughts on waking up are that I can feel the inflammation in my body. I have reduced prednisone for two days and I can feel the effects. I have the sense that in the dream I was not desperate to just save my own life or I could have thrown the can anywhere. I also cared about the lives of others. I had complete focus and determination to hold the can tightly in *hopes* it would put the fire out. I was *very* relieved to find the fire go out as were the others in the dream. I feel courage and determination even though I feel sick this morning. I am going to assume it is prednisone withdrawal and not inflammation.)

---

[104] Chuck did not indicate the sources he consulted for these two quotations.

This was a crucial dream. As with Ruby, the prednisone that was relieving the inflammation caused by the autoimmune disorder was also threatening his life. Take it too long and it breaks the body down. Consequently, how much to take and when to reduce the dosage are critical questions.

When Chuck reduced the dose and his inflammation flared up, it could have been due either to the reduction in prednisone or to a reemergence of the disorder. If the former, then his best choice would be to tough it out. If the latter, he would need to restore the prednisone to its previous level.

Chuck used this dream as an answer to the question of whether or not to restore the dosage. While that was risky, he felt he could trust it. The next day, his symptoms were worse. But he continued to trust the dream. A few days later the symptoms had abated and he felt much better.

He had reduced the prednisone and his health did not suffer. His trust in the dream had been confirmed.

Another fire dream followed shortly:

> *This Time God Doesn't Need Help*
> I am sitting next to God. God is holding a container that is covered with cloth. The outside of the cloth has a small fire on it. God is trying to put it out. I try to help and he motions me away—shoos me away.
> I somehow understand that God started the fire and does not want or need my help putting it out.
> (This is a very vivid dream. Upon waking I am at first concerned about the symbol of fire in my dream. Reliably, fire has been a symbol of "inflammation" in my body. Then I realize that God is the one that started this fire and doesn't want my help putting it out. At least not now, not in this way. Then it becomes comforting to realize that God is holding it and is trying to put it out.)

To this day Chuck considers this dream to be the most significant he has had. That God doesn't need his help tells him that God is in control. Chuck can let go of the reins of his disease. He can trust that forces more powerful and insightful than him are dealing with the disorder. No matter what the outcome, he will be taken care of. The dream was also the confirmation he had been looking for.

His decision to trust his dreams to indicate when he should intervene or not in his inflammation was paying off. He now felt he had a reliable indicator of how to approach the disorder. He could apply himself to improving his life rather than to coping with the disorder.

He cleaned up debris. He planted a garden. He cultivated native plants. He watched as the life in the land on which he lived flourished. He began to feel like his old self.

The autoimmune disorder did not go away. He kept his appointments with his practitioners faithfully. He continued to learn as much about the disorder as he could. He engaged his practitioners in dialogues about what he had learned. He began to understand the frustration they felt in not being able to give him the time he deserved, as his insurance company would reimburse only short-time segments.

That gave him a great deal of sympathy with his practitioners. He was becoming an empathic member of the world around him: His land, his garden and his practitioners. He was not simply feeling like his old self. He was recovering his old self.

As of this writing, Chuck continues to do well. He still has an autoimmune disorder and the future is not at all certain. But he is living vitally in the present.

His most recent dream was:

> A brief snapshot type dream. I am standing in a forest with another person next to me. In the distance I can see smoke billowing up above the tree line from a house that is on fire. It is my house. Somehow I sense reassurance from the person next to me that everything is going to be OK.
>
> (I wake up from this dream quite concerned. Dreams about my house frequently have to do with my physical body and dreams about fire are usually about inflammation caused by the autoimmune disorder. Recently I have not been doing well on my prednisone reduction. Not terrible. But not good either. The dream confuses me because typically fire is an indication that I am in for a rough ride and yet I keep going back to the feeling of reassurance given me within the dream.
>
> I decide to interpret the dream that it is in part giving me validation that, yes, I am not doing well. The inflammation could be from the disorder or possibly from some other bodily cause.
>
> If from some other cause, it may be transitory. If from the disorder, it

seems possible that I am being told that staying on my current dose I will be OK. This could be because of some of my other medications for the disorder are possibly finding new effectiveness. The other drugs are slow to change the course of the disorder—hence the use of prednisone to buy time. It seems reasonable to my conscious mind to stay at my current dose of prednisone and wait and see if I improve or remain the same. I will stay alert that if I get worse I will adjust my prednisone quickly.

Update—a week later: It appears that my interpretation was right. If not right, was at least not wrong. I did not get worse as the house on fire symbol alone would have led me to believe.)

Chuck lives his life waking as he lives his life dreaming. In a rather seamless way he has integrated conscious and unconscious. That reassures him and provides new confidence in his unconscious's ability to inform him of the state of his body.

Where previously he was hypervigilant with regard to the disorder, spending hours a day researching it on the internet, now he could give it its proper place in his life. Where previously his preoccupation with his disorder crowded virtually everything else out of his life, now life was opening up to him. He was finding new involvements and inclusions.

Please note that at no time did Chuck abandon his medical treatment. Rather, he developed a healthier relationship with it. Chuck's consistent second-guessing of his practitioners yielded to a more cooperative relationship of mutual research into his disorder.

Never did Chuck replace his medical treatment with his relationship with his unconscious. This was not an either–or proposition. It was both–and.

Conclusion

The point of these two stories is not that psychology can cure physiology. Ample evidence exists, of course, that it sometimes can. But that is a decision that cannot be made during the course of an illness. One should explore all forms of healing. Rarely does only one suffice.

That said, one can only wonder about the role of psychology in autoimmune disorders. One thinks of the relationship between anger and inflammation, sorrow and edema, worry and ulcers, heartache and coronary disease. To discount the psychological factors would be ridiculous.

Likewise to forswear medical treatments.

Comprehensive approaches are usually the best course of action. For our purpose, it is important to note that most psychological approaches focus on consciousness. That leaves comprehensive approaches vastly deficient, for if medical treatment and psychological techniques leave out the unconscious, the primary interface between psyche and soma is simply ignored.

Thus it is that the psychoid realm of the psyche is essential for healing. Not exclusively, of course. But essential. Both Ruby and Chuck were greatly assisted by their openness and creativity in accommodating the psychoid realm in their healing. It transformed their attitude toward their disorders. Attitude, as healing arts throughout the ages can testify, is an important element in any healing.

9
Active Imagination as Waking-World Activity

I never came upon any of my discoveries through the process of rational thinking.—Albert Einstein.

The previous case studies demonstrate how to integrate conscious and unconscious through the use of unconscious imagery, particularly dreams. Active imagination in those instances involves reentering dream imagery and narratives to see how they unfold. That is not the only way to do active imagination, however. As we saw earlier, any form of bringing conscious and unconscious into relationship with each other constitutes active imagination.

The two case studies in this chapter differ from the former, which focused on employing active imagination as an inner, reflective process. We shall now see how active imagination can be lived out consciously in daily life. In both cases, the individuals engaged unconscious dynamics through activities in the waking world.

Tony

Tony entered analysis because he was lonely. He had no family or friends. He lived alone and worked alone. His intelligence earned him a good living in a very challenging occupation that required little interaction with others. He was proud of his accomplishments and could cite them easily and elaborately. Affectively, however, he was depressed. He wasn't sure that he wanted companionship, but he felt lonely without it.

We spent the first sessions getting to know each other. I found him to be quiet and withdrawn, yet a considerate man whose social skills seemed underdeveloped for his age (around forty). But he was cooperative and engaged with the work. Dreams were not a very important part of his life, nor would they become so in the course of the analysis.

The first stumbling-block occurred early on. His hygiene was questionable, if not problematic. He had conspicuous body odor.

For whatever reasons, I have often been uncomfortable commenting on people's lapses in persona. For example, I used to go through periods of ambivalence in social gatherings when a man's fly was open. Should I tell him? Will he be embarrassed? What if I don't tell him? With age I have become less concerned with those kinds of questions as I am approaching the time when my own lapses in attention lead me to appreciate input from others.

Leaving Tony unaware of a behavior that might either alienate him from others or make him the butt of jokes and gossip could hardly be considered helpful. So one day at the beginning of a session I mentioned that his sweater had an odor. I asked him if maybe it needed washing.

He said that he hadn't noticed. But he wondered if that might explain why someone had put a stick of deodorant on his desk at work. From that moment on, his hygiene improved dramatically.

Perhaps because we had both shared a potentially embarrassing moment, he began to share more deeply. He brought in a couple of dreams. He commented on my attire. He told me that his father died when he was five. His stepfather never liked him. He had had a sister, but she died when he was thirteen.

One day, while we were discussing his depression, I asked him if there was anything in life that he really wanted to do. He was silent for a moment and then said: "I want to fly to *X*, buy a ten-year-old girl, and have sex with her."

At first I thought that he was joking. Nothing in his demeanor, however, suggested that. He was serious. I panicked.

What could I do? Nothing in my training—either my PhD or my analytic training—had prepared me for this. On the one hand, I should do all that I could to stop him. On the other, if I pursued that line too forcefully he was likely to leave analysis and never return. Any opportunity to prevent disaster would be lost.

As long as he kept returning, the possibility existed that we could work this through. It reminded me of a colleague in a church where we had conducted the Sunday family Eucharist.

Many of the children who attended were not allowed in their homes on Sunday, for their mothers were prostitutes or drug addicts and didn't want them around.

They were a real challenge. They came to our service for the breakfast we served and for a warm place to escape the cold weather. While they were not particularly interested in the Eucharist, they did form relationships with the Sunday School teachers, their classmates, my colleague and me.

My colleague was a Bahamian who had worked as a fishing guide. One day he told me about a time when he had helped a fisherman to catch the largest barracuda ever caught on one-half pound test line. I was incredulous. "How much did it weigh?" "Forty-five pounds," he replied. "How can that be possible? The line breaks when more than a half-pound is put on it."

"Well," he replied. "When you hook the fish it swims away as fast as it can. You have to keep letting out line as long as it swims away. When it stops, you gently begin to reel it in. Of course, then it takes off again and you have to let out more line. Eventually the fish is exhausted. Then, you can gently reel it in and net it at the boat."

"How long did it take?" I asked. "About eight hours," he replied. "How much line did you have out?" "About a mile." "A mile?!" "Yes, that's what it took."

We both realized that that was a pretty good metaphor for working with the children who came to us off the street. Years later, when I was working with Tony and discovered his life's dream to fly off to another country, buy a ten-year-old girl and have sex with her, it again became an appropriate metaphor. I was fishing for barracuda with half-pound test.

Tony had told me that his father died when he was five and his sister had died when he was thirteen. He also revealed through the course of our work that his sister was the only family member with whom he had a close relationship.

His mother was entirely too seductive for him to open himself to her. His sister was all he had. In fact, when she died he tried to kill himself by sitting in a puddle of water in the rain and catching pneumonia. To his disappointment, he had been taken to the hospital where, although it was nip and tuck for a while, he survived.

He had been depressed ever since. That was not surprising. Then it occurred to me to ask him how old his sister had been when she died.

"Ten," he answered.

What a shock! Thoughts arranged themselves in my mind like figures on a slot machine. Had we hit the jackpot? Could he be looking for his sister in the ten-year-old girl he wanted to buy? Was he longing to recover the only intimate relationship he had ever had, a relationship that was torn away from him when he was thirteen? Had he longed for a ten-year-old girl ever since?

We were about to enter his fantasy in a way that (I hoped) would generate bifurcations and oscillations that could lead to a chaotic transformation. I asked him if there was any relation between his sister dying at ten and his desire for a ten-year-old girl.

He began to oscillate. His mind seemed to race as it oscillated between sister and desired girl-love-object. He didn't know what to say. At that point, there really wasn't anything to say.

He was torn between his beloved fantasy and its concrete reality.

We needed to let out some line. We dropped the subject and for the next several sessions chattered away on less volatile subjects: His job, relations with other tenants in his rooming house, and so on.

It soon became time to construct another bifurcation to elaborate his oscillation. One day, as we were talking about his sister, I asked him: "How do you think you would have felt if your parents had sold your sister to a man for sex?" It was a risky question, but it had the needed impact.

He turned bright red. His depression now flared into anger. He was furious with me for posing the question. But then he realized he was furious with himself for wanting to do to another what would have been horrifying to have done to his sister. Spiraling into despair, he began to cry.

He needed another bifurcation. I told him that in the country where he hoped to purchase the ten-year-old girl there were several charitable organizations that were trying to rescue young girls from sexual slavery. Perhaps he could donate the money he was saving for his airline ticket.

He was silent. He had had enough for one day. He left without comment at the end of the session. At the beginning of the next session he said he had made a decision. He had researched the charities, found one he particularly liked, and decided to donate the money he had saved for his airline ticket to that charity. On the one hand, he was relieved. On the other, he was sad at the demise of his fantasy. His connection to his sister would not be renewed.

Or maybe it could. We needed a solution that would feel good and would connect him with his sister now that Plan A was no longer in play. The unconscious offered up a solution. It suddenly appeared in our shared oscillations in both of our minds essentially simultaneously: He would donate it in her name. He would donate the money in memory of his sister.

Now we both felt like crying. My tears were of relief. His were of finally connecting in a meaningful way with the sister he had lost. In that moment we were both fully in the realm of the unconscious. We never could have figured this out cognitively. Even worse, any attempt to restructure his cognition to eradicate the fantasy would put him to flight. He didn't need correction. He needed his sister.

Active imagination continued throughout his implementation of his plan. At every step—researching the charity, deciding which one to support, writing the check, sending it off in the mail—he was connecting with the remnants of his sister that called to him from his unconscious. They were together again. The connection was repaired. His desire to be intimate with a ten-year-old girl had been fulfilled.

All the thinking that we had done about his fantasy had involved and included active imagination. We had engaged the fantasy and considered alternatives, bifurcating thereby a deterministically chaotic dynamic in the unconscious. The fractal attractor generated by his childhood trauma had been transformed into an entirely new one. The past disaster had been redeemed.

Active imagination as it plays out in our daily activities can take many forms. The next example shows how spontaneous and simple the process can be.

Roy

Roy was referred to me by a colleague for his depression. Two months earlier he had become ill, which left him with "no energy, needing rest, feeling ill at ease, enervated, experiencing imbalance, difficulty focusing, frustrated and 'down'." As his blood tests were normal, his physician recommended anti-depressants. He decided instead to try counseling, for which he consulted my colleague. Never before had he tried psychotherapy or analysis.

As he had read *Memories, Dreams, Reflections* and was interested in Jung, he willingly recorded his dreams. Over time we analyzed many of them and did active imagination by engaging dream figures in imaginal dialogues.[105] He wrote these down in order to polish them afterward.

Interactions with dream imagery energized him. He discovered much about himself, his family, and his life history. Meanwhile, he was promoted to a higher position at his job. As he adjusted to the new environment he discovered an enthusiasm for his work that led him to develop creative solutions to perennial problems. His boss was very pleased.

Earlier, he had given this creative enthusiasm free reign. He traveled, explored, sought out new friends and experiences, and lived an energetic lifestyle. Over time, however, he realized that he was not putting down any roots. He was spontaneous, but not centered. He felt that he was experiencing life broadly, but not deeply. A few years before he sought analysis, he decided to settle down.

Could that have been a source of his depression? Was he bored with life? Did he feel that he was missing out on new experiences? If so, the analysis seemed to be filling the void. He applied himself to exploring and engaging the imagery that occurred in his dreams.

The following dream gave us something to work on:

> I was in a vast underground storage area carved out of sandstone. There's lots of sunlight shimmering against the walls, but I can't see the windows. The room and hallway are like a warehouse, and they're all but empty. There are a few boxes with odd items resting on them. All I remember is the plaster cast of a shark's jaws and teeth. I can't tell if I'm moving things in or out, but they're staying put for now.

---

[105] *Imaginal* differs from *imaginary* with regard to the reality of the phenomenon. *Imaginary* is often thought of as *unreal*, as in *imaginary pain*. *Imaginal* refers to the images the psyche produces. These are very real, as in Jung's phrase: "Psychic reality is reality."

While the dream activated memories of Roy's former lifestyle, his main interest was the shark's jaws. He had seen them in shops before, but never had they caught his attention as vividly as in this dream.

We looked at how sharks function. They must always be in motion because their gills are not movable. For water to flow over their gills, sharks must be moving. Voracious in their feeding habits, they ingest whatever is around them. Consequently, their stomachs become loaded with everything they have ingested: bones, trash, flotsam and jetsam.

He began to wonder if his appetite for experience could sometimes resemble a shark in motion. Does he rush too headlong into things? Does he absorb too much of what is around him? The image of the shark's jaws could be a template by which to evaluate life experience. He decided to keep his eyes open for the appearance of shark's jaws or anything that resembled them: Active imagination in the waking world, so to speak.

On his way home from the session, he passed a store that was silk-screening tee-shirts. Amazingly, one was all black except for the white image of a shark's jaws. Struck by the image, he bought it immediately.

He related this story at the beginning of our next session. He described how he had worn the shirt to see how it may have reflected what was going on in the environment. With some experiences it resonated. With others, not so much. He found that it comprised a commentary on his life experience.

In itself, that was a very informative active imagination. But there was more. The unconscious seems to have had more in mind. As we played with the image in our session, it suddenly occurred to us that the shirt itself had something to say. He could put it on or take it off at will. What a great metaphor! Could he also put on or take off his shark's-jaws' orientation to life? Could he decide when to give it free rein or when to rein it in?

He loved the idea. In his daily experience he could dialogue with the image of the shark's jaws to discover what might be the best approach to whatever presented itself. The single-event active imagination of purchasing the tee-shirt evolved into an ongoing active imagination with regard to life itself.

The more he practiced this form of active imagination, the more he noticed how others acted out their own shark's-jaws' orientations to life experience. It became a shorthand way of discerning unconscious dynamics of interpersonal relationships. Conscious and unconscious shared a view of human life. It was exhilarating and enlivening.

Energized as he now was by the act of opening conscious to unconscious, he began to take seriously the thoughts and ideas that occurred to him. He had purchased a house and thought about planting a garden. Recognizing an opportunity for active imagination, he set about preparing the ground for planting. In consideration of the short growing season in the Pacific Northwest, he set up small greenhouses of tubing and plastic to protect the young plants from the springtime cold.

Due to the shape of his lot, he made six greenhouses on each side of his house. One afternoon, after he had put all the tubing in place, he looked upon his handiwork. He was startled to see that there was a pattern. On both sides of his house sections of straight tubing stuck out at right angles.

He was immediately reminded of the *I Ching* and its hexagrams. His house was in the middle of six *yin* lines, which make up hexagram Two, THE RECEPTIVE, the most feminine of all the hexagrams. What was the earth in his garden, if not pure receptivity awaiting planting?

Roy was stunned. Here was a living image from his unconscious that portrayed the fertility not only of his yard, but of his unconscious as well. He was literally living in the middle of receptivity. His approach to life was illustrated for him in the form of his garden.

Could there be a more vivid example of active imagination? His yard symbolized his unconscious, which was prepared to offer all that the conscious needed. What existed in his psyche, existed in his land. Each reflected the other.

Summary

With this last example we come to the end of our elaborations of active imagination. Whether reintegrating split-off parts of our psyches or discovering new ways of perceiving our lives, conscious immersion in unconscious is always fruitful. And, for good or for ill—chaotic.

Thus it is that when we begin to practice active imagination we should do so under the guidance of someone who has extensive experience, both personally and with others. Immersion in the unconscious is not to be taken lightly. We can easily lose our way. That which is symbolic must be taken literally only with great reserve. We should never do active imagination when our egos and corresponding consciousness are impaired.

In our final chapter we'll review what we have discovered. It's been a long journey. Hopefully, landmarks and navigational aids have been sufficiently elaborated to give us some assurance that the journey can be useful.

That, of course, is a personal decision. But for those of us who have already embarked, we welcome the company of those who are prepared to begin. To them we say in all sincerity: "Bon voyage!"

10
Onward

It is known that man has the ability to immerse himself entirely in one subject, however insignificant it might seem. And it is known that there is no subject so insignificant that it will not expand to infinity, if attention is concentrated on it.

—Tolstoy.

Relationship is the essence of morphogenesis. It so permeates physiology and psychology that we could say that it is the supreme archetype. By attracting us to that which furthers individuation and repelling us from that which hinders it, relationship is the archetypal dynamic through which all archetypes accomplish their ends.

Everything in this book is about relationship: Between conscious and unconscious, childhood and adulthood, person and person, bifurcating and oscillating system with bifurcating and oscillating system—and perhaps most importantly, author and reader. Disturbances in individuation are caused by inadequate relationships. Healing of those disturbances is possible only through relationships. That between analyst and patient encourages that between patient's conscious and unconscious.

Those simple phrases hardly communicate how difficult it all is. Those who enter analysis are committed to finding a way out of their dilemmas. They are willing to cooperate with whatever will help them to heal. They look to us for insight and guidance.

That has its shadow side, however. When we enter analysis, we might as well be saying: "I'm suffering from something about which I haven't a clue. Although I've lived with myself for *X* many years, I don't know what's wrong. I just feel terrible. Please figure it out. However, because it is so painful, I'll block you at every turn. Not consciously or deliberately. It will just happen."

As I noted in the introduction, it's not rocket science. It's far more difficult. Resistance is rarely deliberate. Conscious and unconscious dwell in totally different worlds. Consciousness inhabits diachronic time while the unconscious lives in synchronic time. Analysis brings conscious into unconscious by surrendering diachronic time to synchronic time, which is why creating dialogues between the two can stretch the imagination beyond the comfort zone of waking life.

That's where active imagination comes in. In active imagination we engage an unconscious image or narrative with no intent whatsoever to make it become anything in particular. Where it will lead, we do not know. We simply dwell in its presence. Consciously and deliberately we join up with the morphogenesis already active in our bodies. We surrender ourselves to that which makes us who we are, yet which we cannot see, analyze, or control.

By allowing morphogenesis to unfold, we engage deterministic chaos by bifurcating with it. That creates a new bifurcating and oscillating system. The old patterns that kept reappearing fractally as endless chaos can now evolve into new patterns as conscious oscillates with unconscious. By engaging our inner child selves, we liberate the determinism of the chaos so that morphogenesis can transform it into new configurations.

Then, the process continues. Oscillations that generate recurring patterns can be further engaged by consciousness—with no ultimate goal whatsoever—to create new oscillating systems. Individuation is freed from the past by integrating it into the present. Thus does analysis transform.

That it all happens through morphogenesis places us in the dark feminine. There, where consciousness holds no sway, miracles occur. Just as a lifeless zygote develops into a living, breathing person, the latent potential for development inherent in our psyches transforms our personalities into new configurations.

It requires us to be both masculine and feminine. The former stands firm, penetrates and fertilizes. The latter receives, contains and incubates. We must forcefully thrust ourselves into a state of being where we can accommodate, receive and incubate the inconceivable. We cannot control what then happens. It takes place outside our volition. We must surrender to the dark feminine trusting that morphogenesis will do us well.

Review

Part One looked at the physiological contributions to morphogenesis. That our bodies generate shape from the moment of conception is a truly miraculous process, one that we barely understand. We then added Jung's explanations of the psychological analog to those physiological processes. We noted that when a childhood environment is so disordered that archetypes have nothing healthy to which they can direct our attention, chaos ensues.

That led us to chaos theory, which explains the complexities of deterministic chaos. From that we learned that a bifurcating system can become stuck in repetitive patterns until another bifurcating system engages it. Childhood defenses fit that pattern. They are paradoxical and cry out for engagement in order to be transformed. Trauma theory gave us some tools that we can use to guide that engagement.

Then we applied what we had learned in Part One to clinical applications in Part Two. We proposed a four-part scaffold by which to identify paradoxical defenses. Next we saw how consciousness exists in diachronic time whereas the unconscious exists in synchronic time.

That helped us to see that those aspects of ourselves that we had thought were left behind (or had never known to begin with) were tied to consciousness's orientation to diachronic time. In the unconscious, however, synchronic time is the eternal *now*. Thus, we could rework the past in the unconscious present.

Active imagination was the topic of the last two chapters. We first looked at autoimmune disorders, which have both psychological and physiological dimensions. The last chapter presented everyday ways of doing active imagination.

With that, our explorations of how to heal in adulthood childhood trauma that never particularized in discrete episodes comes to an end. But while this book is finished, our explorations are not. They will go on forever, for the mysteries of the psyche will never be completely resolved. Our explorations are asymptotic: Always approaching the destination but never getting there.

In analysis, the journey truly is the destination.

Adieu

As with the first volume in this series on archetypes and chaos, this volume intends to be an experience of what it describes. In a way, the entire book is an exercise in active imagination. If, after reading it, you cannot figure out what I am saying, don't blame yourself. It simply means that I am not up to the task of deciphering all this and presenting it in an accessible manner.

But it did engage us in a mutual active imagination. That alone makes it worthwhile. For our parting words, then, let's simply say "Welcome to analysis."

APPENDIX
Three-Breath Relaxation Technique

The following relaxation technique can help you to relax more quickly and easily in whatever situation you find yourself. The goal of the technique is to become aware of how well things can go when you are not in control. You will learn what it feels like to completely let go of your body. You will also experience your bodily functions continuing just fine without any input from you. Here is how to do it:

1) Place yourself in a position in which you can completely relax. Lying down on your back often works best. Feel free to place a pillow behind your head, or under your knees. There is no exclusively correct way to do this. Whatever works best for you is the best way to relax.

2) Feel your body sink into the surface on which you are lying. To facilitate this, quickly take as deep a breath as you can, and then let it out. As your chest collapses, feel all the tension in your body drain into the surface you are lying on. Also, feel the surface holding you up. If your body is uncomfortable, feel free to adjust your position.

3) Wait a moment before you take your next breath. This need not be an uncomfortable wait. Just try not to breathe for a bit.

4) Take another deep breath and let it out. Again, feel your chest sink into your body and your body into the surface you are lying on. Feel every bit of tension drain through you and down into the surface you are lying on.

5) Wait another moment before taking another breath. Notice that your heart is beating, your blood is circulating, your cells are being nourished, life is going on without you having to do anything to sustain it. You don't even have to breathe. Your body is on automatic and you are free from having to do anything at all.

6) When you are ready, take a third deep breath, and let it out. Again, feel your chest sink into your body. Notice that any remaining tension, worry, anxiety, or conflict drains out of your muscles into the surface on which you are lying.

7) As your lungs empty, you will now have sufficient oxygen in your system to allow you to refrain from breathing for quite some time. In this quiet space where you do not have to do anything to stay alive—you don't even have to

breathe—you can feel what it is like to have everything being done for you. Everything is being taken care of for you. There is absolutely nothing you must do to stay alive.

The first time you do this exercise, it may feel a bit strained or awkward. More complete relaxation will come with practice. At first, try to practice three times a day. The goal is to make the relaxation response a conditioned reflex. That is, it becomes automatic when you take a deep breath and let it out. Eventually, you will be able to relax completely without having either to take three deep breaths or to lie down. Simply taking a breath and letting it out will activate the relaxation response.

Glossary

NB: Terms and concepts thoroughly explained in the text are not generally included here. See Index if not here.

amigdala: the part of the brain where, among other functions, perceptions are translated into emotions.

antibodies: a component of the adaptive immune system.

antigens: molecules not of the body that induce immune responses.

autoimmune disorder: a disorder that results from the immune system attacking its own body.

bifurcation: a split in a dynamic that establishes two dynamics where there was previously only one. If the dynamic continues to bifurcate, then four, eight, sixteen, thirty-two (and so on) dynamics evolve from the original.

biochemistry: the interactions of atoms and molecules that are inextricably intertwined with biological processes. For example, the biochemistry of the brain focuses on the chemistry of neurotransmitters and receptors, which enable the brain to function.

cortisol: a glucocorticoid that, among other things, reduces inflammation.

epigenetics: the chemistry by which DNA is transcribed into RNA and RNA is translated into proteins.

fractal: a term coined by Benoit Mandelbrot, referring to elements of a dynamic that cannot be defined by categories.

frontal lobe: the part of the brain just behind the forehead. It comprises the higher cognitive functions, such as impulse control.

genome: the genetic material (DNA organized into genes, which are bundled into chromosomes) found in every cell.

DNA: deoxyribonucleic acid. It is what genes are made of.

RNA: ribonucleic acid. One strand of the double helix DNA. RNA serves as the template for generating proteins.

prions: infectious proteins that do not fold as expected.

proteins: genetic material that, when folded, forms the stuff and the mechanics of the body.

glucocorticoid: a steroid that, among other things, reduces inflammation.

heliophobic: repelled by the sun.

heliotropic: attracted to the sun.

hippocampus: the part of the brain where, among other things, memories are formed and stored.

HPA axis: the interactions among the hypothalamus, pituitary and adrenal glands. The HPA axis plays a role in the production of glucocorticoids, which influence how the body responds to stress.

immune system: our body's system designed to combat threats to its existence.

adaptive immune system: the part recognizes specific pathogens through previous experience encountering them. It is our second line of defense.

innate immune system: the part that does not require experience to recognize pathogens, and recognizes anything that appears to be foreign. It is our first line of defense.

immunology: the study of the immune system.

inflammation: an immune response that may cause a part of the body to feel like it's on fire. It is often manifested as redness and swelling.

limbic system: the part of the brain that is involved with instincts, drives, and emotions. It often manifests as the "fight or flight" response to threat.

metapsychology: the basic concepts upon which a psychological school of thought is based.

microbiology: the study of small organisms and genetic material, including bacteria, viruses, prions, fungi, parasites.

microclimate: an environment that exists within another environment. For example, the temperature and humidity of the environment under a canopy of trees or grass may be very different from that of the environment where the trees or grass are located.

morphogenesis: the generation of shape, continuous from birth to death.

neuroscience: the study of the nervous system, particularly that of the brain.

neurotransmitters: chemicals that flow through the synaptic cleft and are taken up by receptors in the receiving neuron. Some of the more commonly recognized are seratonin and dopamine.

oscillation: the switching back and forth between states, for example a pendulum that swings back and forth, or an alternating current in which the flow of electrons regularly and consistently reverses direction.

parasites: organisms whose existence depends on taking advantage of the resources of another organism.

pathogens: substances foreign to the body that can do it harm.

plasticity: the ability for something to change. *Synaptic plasticity* refers to the generation and alteration of synapses.

prednisone: a glucocorticoid to reduce inflammation.

psychoid: the mysterious realm of the psyche where body and mind overlap.

regression: going backwards. When as adults we react to a situation with emotions appropriate to childhood, we have *regressed*.

scale-invariance: refers to patterns that preserve their appearance no matter how they are magnified or reduced. In other words, they look the same no matter what the scale.

schizophrenia: the "disease with a thousand faces," which generally refers to lack of ego and to psychotic thinking. Contemporary theory attributes the condition primarily to genetics, although there is considerable evidence that it can also be caused by environmental influences.

self-similarity across scale: the same as scale-invariance.

sequelae: the effects of a phenomenon. For example, coughing is a sequela of tuberculosis.

synapse: the part of a neuron that communicates with other neurons.

synaptic cleft: the space between synapses, across which neurotransmitters flow, enabling neurons to communicate with each other.

synaptic configuration: the biochemical arrangement of a synapse that allows it to produce consistent results with similar inputs from other synapses.

theriomorphic image: one that consists of two or more elements combined, for example, Quetzecoatl, which is a winged serpent.

transposons: segments of genetic material that wander the genome until they find a place to insert themselves, thereby altering the genome.

BIBLIOGRAPHY

Cacioppo, John T. and Decety, Jean. "What are the Brain Mechanisms on which Psychological Processes are Based?" In *Perspectives on Psychological Science,* vol. 4, no. 1 (January 2009).

Conrad, Joseph. *Lord Jim.* New York: Modern Library, 1931.

Diener, Ed. "Neuroimaging: Voodoo, New Phrenology, or Scientific Breakthrough? Introduction to Special Section on fMRI." In *Perspectives on Psychological Science*, vol. 5, no. 6 (November 2010).

Dorner, Dietrich. Trans., Rita and Robert Kinber. *The Logic of Failure: Recognizing and Avoiding Error in Complex Situations.* Reading, MA: Perseus Books, 1996.

Dostoevsky, Fyodor. *The Brothers Karamazov.* Trans., Richard Pevear and Larissa Volokhonsky. New York: Farrar, Straus, and Giroux, 1990.

Franzen, Jonathan. *Freedom: A Novel.* New York: Farrar, Straus, and Giroux, 2010.

Freeman, Marc R. "Specification and Morphogenesis of Astrocytes." In *Science*, vol. 330, no. 6005 (November 5, 2010).

Freud, Sigmund. *Therapy and Technique.* New York: Collier Books, 1963.

Goldberger, Ari L., Rigney, David R., and West, Bruce J. "Chaos and Fractals in Human Physiology." In *Scientific American,* vol. 262, no. 2 (February 1990).

Hemby, S & O'Connor, J. "Transcriptional Regulation in Schizophrenia." In Romer, D. and Walker, E., eds., *Adolescent Psychopathology and the Developing Brain: Integrating Brain and Prevention Science.* New York: Oxford University Press, 2007.

Hulbert, Ann. "Review of *A Widow's Story* by Joyce Carol Oates." In *The New York Times Book Review* (February 20, 2011).

Jung, C. G. *The Collected Works of C. G. Jung* (Bollingen Series XX). 20 vols. Trans. R. F. C. Hull. Eds. H. Read, M. Fordham, G. Adler, Wm. McGuire. Princeton: Princeton University Press, 1953-1979.

__________. *Memories, Dreams, Reflections.* Ed. Aniela Jaffé. New York: Pantheon Books, 1961.

Jung, Emma. *Animus and Anima.* Dallas, TX: Spring Publications, 1981.

Kepler, Thomas B., Marder, Eve, and Abbott, L.F. "The Effect of Electrical Coupling on the Frequency of Model Neuronal Oscillators." In *Science,* vol. 248 (April 1990).

Kesey, Ken. *Sometimes a Great Notion.* New York: Viking Penguin Inc., 1964.

Kundera, Milan. *Life is Elsewhere.* New York: HarperCollins, 2000.

LeDoux, Joseph. *Synaptic Self: How Our Brains Become Who We Are.* New York: Penguin Books, 2003.

Mampe, Friederic, and Wermke, Christophe. "Newborns' Cry Melody is Shaped by Their Native Language." In *Current Biology,* vol. 19, no. 23 (November 2009).

Mandler, Jean M. "How to Build a Baby: II. Conceptual Primitives." In *Psychological Review,* vol. 99, no. 4 (October 1992).

Poldrack, Russell A. "Mapping Mental Function to Brain Structure: How Can Cognitive Neuroimaging Succeed?" In *Perspectives on Psychological Science.* vol. 5. no 6 (November 2010).

Razzell, W. and Martin, P. *Embryonic Clutch Control.* In *Science,* vol. 335 (March 2012).

Romer, D. and Walker, E. *Adolescent Psychopathology and the Developing Brain: Integrating Brain and Prevention Science.* New York: Oxford University Press, 2007.

Samuels, Andrew. *Jung and the Post-Jungians.* New York: Routledge and Kegan Paul, 1986.

Sharp, Daryl. *C. G. Jung Lexicon.* Toronto: Inner City Books, 1991.

Sullivan, Harry Stack. *The Interpersonal Theory of Psychiatry.* New York:

W. W. Norton & Co., 1953.

Turner, Samuel M. and Herren, Michel, eds. *Adult Psychopathology and Diagnosis.* New York: John Wiley and Sons, 1984.

Van Eenwyk, John. "The Analysis of Defenses." In *The Journal of Analytical Psychology,* vol. 36, no. 2. (April 1991).

__________. *Archetypes & Strange Attractors: The Chaotic World of Symbols.* Toronto: Inner City Books, 1997.

__________. "Switching Tracks: Parallel Paradigms in Psychology and Religion." In Shafranske, Edward P., ed., *Religion and the Clinical Practice of Psychology.* Washington, DC: The American Psychological Association, 1996.

Zimmer, Carl. *Parasite Rex: Inside The Bizarre World of Nature's Most Dangerous Creatures.* New York: Touchstone Books, 2000.

Index

Studies in Jungian Psychology by Jungian Analysts

Quality Paperbacks

Prices and payment in $US (except in Canada and Visa orders, $Cdn)

Bees, Honey and the Hive: Circumambulating the Centre.
Frith Luton (Melbourne) ISBN 978-1-894574-32-7. 208 pp. $30

Risky Business: Environmental Disasters and the Nature Archetype
Stephen Foster (Boulder, CO) ISBN 978-0-919123-33-4. 128 pp. $25

Miles To Go Before I Sleep: Growing Up Puer (a Jungian Romance)
Daryl Sharp (Toronto) 978-1-894574-36-5 128 pp. $25

Clinical Chaos: The Strange Attractors of Childhood Trauma.
John R. Van Eenwyk (Olympia, WA) ISBN 978-1-894574-37-2. 192 pp. $30.

The Sacred Psyche: A Psychological Approach to the Psalms
Edward F. Edinger (Los Angeles) ISBN 978-1-894574-09-9. 160 pp. $25

Eros and Pathos: Shades of Love and Suffering
Aldo Carotenuto (Rome) ISBN 978- 0-919123-39-7. 144 pp. $25

Descent to the Goddess: A Way of Initiation for Women
Sylvia Brinton Perera (New York) ISBN 978-0-919123-05-2. 112 pp. $25

The Illness That We Are: A Jungian Critique of Christianity
John P. Dourley (Ottawa) ISBN 978-0-919123-16-8. 128 pp. $25

Coming To Age: The Croning Years and Late-Life Transformation
Jane R. Prétat (Providence) ISBN 978-0-919123-63-2. 144 pp. $25

Jungian Dream Interpretation: A Handbook of Theory and Practice
James A. Hall, M.D. (Dallas) ISBN 978-0-919123-12-0. 128 pp. $25

Phallos: Sacred Image of the Masculine
Eugene Monick (Scranton) ISBN 978-0-919123-26-7. 30 illustrations. 144 pp. $25

The Sacred Prostitute: Eternal Aspect of the Feminine
Nancy Qualls-Corbett (Birmingham) ISBN 978-0-919123-31-1. Illustrated. 176 pp. $30

Jung Uncorked: Rare Vintages from the Cellar of Analytical Psychology
Three vols. *Daryl Sharp (Toronto)* ISBN 978-1-894574-21-1/22-8/24-2. 128 pp. $25 each

The Pregnant Virgin: A Process of Psychological Development
Marion Woodman (Toronto) ISBN 978-0-919123-20-5. Illustrated. 208 pp. $30pb/$35hc

Discounts: any 1-9 books, 20%; 10-19, 25%; 20 or more, 40% .

Add Postage/Handling: 1-2 books, $6 surface ($10 air); 3-4 books, $12 surface ($16 air); 5-9 books, $16 surface ($25 air); 10 or more, $16 surface ($30 air)

VISA credit cards accepted. Toll-free: Tel. 1-888-927-0355; Fax 1-888=924-1814.

INNER CITY BOOKS, Box 1271, Station Q, Toronto, ON M4T 2P4, Canada
Tel. (416) 927-0355 / Fax (416) 924-1814 / booksales@innercitybooks.net